Geosciences: The Future

Final Report of the IUGG Working Group Geosciences: The Future

Geosciences: The Future

S. Adlen, IAMAS
Atmospheric, Oceanic and Planetary Physics
Clarendon Laboratory
Parks Road
Oxford OX1 3PU
(sadlen@atm.ox.ac.uk)

T. Oki, IAHS
Research Institute for Humanity and Nature
Kyoto 602-0878, Japan
(oki@chikyu.ac.jp)

L. Sánchez, IAG
Instituto Geografico Agustin Codazzi
Division de Geodesia
Bogota - Colombia
(lsanchez@igac.gov.co)

K. Yoshizawa, IASPEI
Division of Earth and Planetary Sciences
Graduate School of Science, Hokkaido University
Sapporo, 060-0810, Japan
(kazu@ep.sci.hokudai.ac.jp)

E. E. Brodsky, IAVCEI, Chair
Dept. of Earth and Space Science
University of California, Los Angeles
Los Angeles, CA, USA
(brodsky@ess.ucla.edu)

A.J. Ridley, IAGA, Editor
Center for Space Environment Modeling
University of Michigan
Ann Arbor, MI, USA
(ridley@umich.edu)

C. Simionato, IAPSO
Centro de Investigaciones del Mar y la Atmósfera
(CIMA/CONICET-UBA)
Dpto. de Ciencias de la Atmósfera y los Océanos
de la Universidad de Buenos Aires
Ciudad Universitaria Pab. II Piso 2
Ciudad Autónoma de Buenos Aires - Argentina
(claudias@at.fcen.uba.ar)

U. Shamir, IUGG Vice President, WG Coordinator
Stephen and Nancy Grand Water Research Institute
Technion - Israel Institute of Technology
Haifa 32000, Israel
(shamir@tx.technion.ac.il)

July 7, 2003

Preface

"IUGG is a purely scientific organisation established to promote and co-ordinate studies of physical, chemical, and mathematical properties of the Earth and its environment in space. These studies include the shape of the Earth; the nature of its gravitational and magnetic fields; the dynamics of the Earth as a whole and of its component parts; the Earth's internal structure, composition, and tectonics; the generation of magmas; volcanism and rock formation; the hydrological cycle, including snow and ice; the physics and chemistry of the oceans; the atmosphere, ionosphere, magnetosphere and solar-terrestrial relations; and analogous problems associated with the Moon and other planets. IUGG activities embrace studies of the Earth by artificial satellites and other techniques for deploying instruments at high altitude." *(Objectives and Activities of IUGG, in the IUGG 2003 Yearbook)*

In summer 2001, the Executive Committee of IUGG (Bureau, Presidents and Secretaries General of the seven Associations) approved a proposal I put forward, to set up a Working Group of young geo-scientists, one from each of the seven Associations, under the title "Geosciences: The Future" (WG-GTF), to examine the state of the geosciences and identify research challenges, emphasizing inter-disciplinary research, integration of scientists from around the world, in particular in less developed nations, and the role that IUGG should play in advancing the geosciences into the future.

The Executive Committee of IUGG, acting as Program Committee for the 2003 IUGG General Assembly in Sapporo, also accepted the proposal to include in the program Union Symposium U8: "Geosciences: The Future," to be prepared and presented by the Working Group.

The WG-GTF, after preliminary discussion and consideration of its expected role and mode of operation, chose its Chair and Editor. Each member prepared a presentation of her/his discipline/Association, sometimes based on material produced by similar initiatives of the Association and/or on responses elicited from other members of their Association. These were circulated among the members, and were discussed in extended conference-calls. Special emphasis was placed on identifying common and inter-disciplinary interests and on encompassing scientists from all parts of the world.

This publication comprises the material prepared by the WG-GTF ahead of the IUGG GA 2003 in Sapporo.

Uri Shamir
IUGG Vice President
Coordinator of the WG-GTF

Contents

Chapter 1

Introduction

E. E. Brodsky

It is both the tradition and responsibility of geoscience to help humanity. Seismologists and volcanologists attempt to save lives by evacuating before eruptions or mitigating the effects of earthquakes. Geodesists quantify the shape of the Earth so that cities can be built and airplanes flown. Space physicists provide a window into the nearly unknown world in which our satellites are sent and spacecraft launched. Hydrologists, oceanographers and atmospheric scientists may have the heaviest burden of all as they help guide the planet through the current climate crisis.

Geoscience also has the responsibility to not confine itself to addressing current problems. As a science, we must develop a framework to address future issues. In order to push knowledge forward, we must invest intellectually and financially in basic science at a level commensurate with the applied fields. Such an investment is intrinsically risky. Many scientific leads turn out to be dead-ends. Other pathways may continue indefinitely, but are difficult to connect to the rest of the field. A field driven by pure thirst for knowledge runs the danger of becoming a tangled spider web of seemingly unrelated facts and personal theories.

Basic science needs to be guided by a long-range vision of the future of the field. Then geoscientists can advance knowledge in a focused way and real progress can be made. The purpose of this report is to present a particular vision that could guide research in the major fields of geophysics and geodesy. This vision is articulated by a young generation of geoscientists with a stake in the future. Together, we have compiled a view encompassing geosciences from the study of the Earth's core to the workings of the Sun.

The main body of this report is organized into chapters representing the domain of each of the associations of IUGG in order from the center of the Earth. Each chapter discusses a long-term vision for the field for the next 50–100 years as well as short-term goals and strategies for achievement. Cross-societal chapters follow the disciplinary section. They outline important directions for interdisciplinary studies, societal benefits and projects targeting less developed countries.

The overall conclusions of the report provide a concise summary of our recommendations for geoscientists in general and IUGG specifically.

This report would not have been possible without the assistance of numerous colleagues who gave generously of their time and insights. As young geoscientists, we had to combine our new point of view with an already extensive base of knowledge and planning in the field. This report is a hybrid of our enthusiastic dreaming and the somewhat more grounded reality of many reviewers, advisers and colleagues. Perhaps together we can see the future.

Chapter 2

International Association of Seismology and Physics of the Earth's Interior

K. Yoshizawa

2.1 Introduction

The primary objectives of IASPEI (International Association of Seismology and Physics of the Earth's Interior) are to promote scientific studies related to earthquakes and physical properties of the Earth's interior, and to facilitate international research and education programs in seismology and physics of the Earth's interior. IASPEI has placed major research emphasis on understanding the nature of earthquakes, the propagation of seismic waves, the physical makeup and dynamics of the Earth's interior, and reducing catastrophic damages caused by large earthquakes.

In this chapter, future research under the umbrella of IASPEI are summarised. We first look at long-term goals for the future development of geosciences relevant to IASPEI. Subsequently, we summarise new challenging topics of existing disciplines covered by IASPEI, which should be tackled in the coming decade, followed by some recommendations for IASPEI and opportunities for interdisciplinary research.

A summary of current research programs of IASPEI has been given in a brochure which is available from an official web site (http://www.iaspei.org/brochure/brochure.html). A comprehensive summary of the development in earthquake and engineering seismology during the last century has been published in *International Handbook of Earthquake and Engineering Seismology* (Lee et al., 2002, 2003). The practice of observational seismology has been summarised in *IASPEI Manual of Seismological Observatory Practice* (http://www.seismo.com/msop/msop_intro.html).

2.2 Long-term Goals

The IASPEI has facilitated researches in understanding physical phenomena taking place in the solid Earth. The study of the Earth's interior is a naturally multidisciplinary subject because we need a variety of information from diverse disciplines, such as seismology, mineral physics, geodynamics and geochemistry, in order to better understand the inaccessible parts of the Earth. In the future, more comprehensive interdisciplinary researches on the close interactions with other parts of the Earth, i.e., oceans and atmosphere, will be inevitable.

Scientific studies of earthquakes are critical to people who live in seismically active areas. The possibility of a short-term prediction of earthquakes is still a controversial issue in seismology. However, the quest for the development of practical methods to predict earthquakes in both the long and the short term, as well as to mitigate catastrophic damages, is essential to reduce threats of earthquakes on our society.

In seismology, a larger class of earthquake phenomenology is still not understandable within the present framework of knowledge, such as the dynamic rupture processes of earthquake faults, high-frequency behaviour of seismic waves in 3-D structures, and the physics of deep-focus earthquakes. Tackling such unsolved problems will require a long-term efforts to develop systematic theories complemented by geophysical observations and laboratory experiments. The accumulation of seismic data over a long period of time will enable us to propose, test and modify hypotheses related to earthquakes and physics of the Earth's interior.

To extend our knowledge on the Earth as a system, including earthquakes, volcanic activities, water circulation, etc., integration of all the available information on the dynamics of the Earth should be pursued.

Towards an Integrated Earth Model

In order to provide a better understanding of the dynamics of the Earth's interior, a variety of information from a wide range of disciplines (e.g., seismology, mineral physics and other geophysical sciences) is necessary.

Seismic waves are one of the most reliable and optimum tools for probing the Earth's interior. They have been utilised not only to investigate the whole Earth structure, but also to explore the natural resources. Of all geophysical methods, seismological investigation provides us with the highest resolution of the inside of the Earth, and thus the development of seismology will, in turn, be the basis to promote the development of solid Earth geophysics.

Combining all available information, such as seismic structure, surface deformation, plate tectonics, numerical results of mantle convection, heat transfer and material circulation in the Earth, we will be able to create a dynamic model of the Earth's interior including both physical and chemical processes in the Earth. In order to interpret the seismic models of the Earth, contribution from other geophysical disciplines, such as mineral physics, geodynamics and geochemistry, are necessary. Furthermore, in the future of geosciences, creating a systematic Earth model, taking account of the interactions between atmosphere, oceans and the solid Earth, will be an ultimate goal to understanding the Earth as a dynamic system.

Water in the Earth: The role of liquid water can be a key to combining the diverse fields of studies related to the Earth's interior, as well as the outer shells of the Earth: ocean and atmosphere. From a perspective of the solid Earth geophysics, it will be most interesting to investigate the presence of liquid water, and its physical and chemical roles in dynamics of the Earth. Recent studies with high-resolution seismic tomography suggest the existence of fluids in fault areas (e.g., Zhao et al., 1996). Further investigation of the role of water in the occurrence of earthquakes will be important for quantitative studies on earthquake mechanisms. Water may also play an important role in the internal dynamics of the Earth, since it has been recognised that a number of mantle minerals are capable of sequestering significant amounts of water in hydrous phases.

Coupling Between the Solid Earth, Atmosphere, Ocean: In the late 1990s, we witnessed exciting new findings about the Earth's continuous tremors even without earthquakes; i.e., the background free oscillations (e.g., Suda et al., 1998; Kobayashi and Nishida, 1998), which have been derived from high quality observations with broad-band seismometers and superconducting gravimeters (Fig 2.1). The cause of this phenomenon is still in the centre of debate, but it is likely to be related deeply to interactions with the Earth's atmosphere and/or oceans. Such interactions of fluid and solid systems will be of significant importance in the coming decade.

The fact that the Earth is vibrating due to interactions with Earth's fluid layers gives us an insight into a possibility of gaining information on planetary interiors without shaking sources in their inside; i.e., it may lead to a basis of the exploration of other planets where there may be no seismic activity but still some atmospheric interaction. Theoretical studies to uncover the mechanisms of the background free oscillations will be essential for applications to planetary exploration.

Planetary Seismology: We can gain many sorts of information on the Earth's interior, but we know very little about the interior of other planets. There are some seismic data from the Moon, and we can deduce a very rough image of its inside, but it is not as clear as what we know about the Earth due to the lack of seismic data. Other planets have been unreachable to seismologists so far, and what we can get is only a rough estimate of their density distribution, which can be estimated from the moment of inertia and mass.

Several projects to explore the Moon and the planet Mars are being conducted. The Japanese LUNAR-A project (Institute of Space and Astronomical Sciences of Japan) will launch seismometers and thermometers using penetrators to study the interior of the Moon, especially focusing on the investigation of the existence of the Moon's iron core. An European and American consortium planned the NetLander mission to bring seismometers to Mars to investigate the Martian interior with a special focus on the existence of buried water and ice, although the mission ran aground due to the withdrawal of two major participating nations.

Through such efforts to bring seismic instruments to the other planets, we will be able to get more concrete information on the deep structure of other terrestrial planets, which will be of great help to better understand the formation and evolution of our planet. A novel technology to bring high-resolution instruments to other planets at a reasonable cost will be useful for the development of planetary seismology. Furthermore, international cooperation and partnership will be essential to make the missions of planetary exploration feasible and successful.

2.3 Short-term Priorities: Issues and Strategies

2.3.1 Seismic Observation and Data Acquisition

Unlike the atmosphere and oceans that cover the solid part of the Earth, the Earth's interior is invisible and untouchable. Information on the inside of the Earth can only be derived from observations near the Earth's surface. Seismic stations play a role as a "telescope" to look into the Earth and provide us

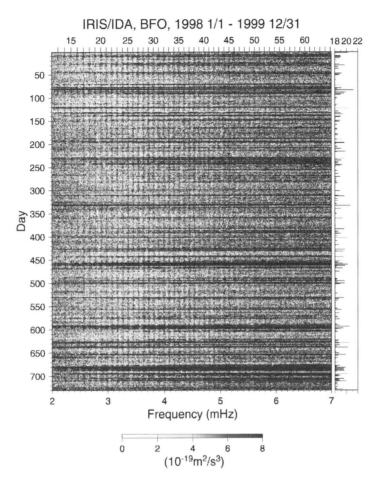

Figure 2.1: A spectrogram for a period between 1998 and 1999 observed at BFO station of IRIS/IDA network showing the Earth's background free oscillations. Labels on the bottom indicate eigenfrequencies of fundamental spheroidal modes, and those on the top indicate angular orders of the modes. A histogram on the right displays seismic moments (Nm) in a logarithmic scale, which indicates the occurrence of large earthquakes during the period of observation. Large amplitude spectra are shown in blue; i.e., the horizontal blue tracks represent prominent amplitudes caused by large earthquakes. We can see vertical traces throughout the period of observation, even on "quiet" days, with no earthquakes (yellowish area), which are extremely well correlated with the angular orders of spheroidal free oscillations. These vertical traces indicate the existence of background free oscillations of the Earth, which are caused by some other sources than earthquakes [courtesy of N. Suda].

with precious information about physical processes of Earthquakes, as well as the structure of the solid Earth. For future development of the studies of earthquakes and Earth's interior, continuous efforts to expand the seismograph network, at both global and regional scales and with high-quality instruments, are essential.

Enhancing Global Coverage of Seismograph Network

Currently, over 120 permanent seismic stations affiliated to FDSN (Federation of Digital Broad-Band Seismic Network) are operated all around the globe, which make it possible to reveal a large-scale ($\sim$ 1000 km) 3-D image of the Earth's inside. Even so, the distribution of such seismic stations is far from uniform, and some regions, such as most parts of the oceans, are yet to be covered. It is critical to achieve uniform distribution of seismic stations with uniform seismometer characteristics to make observations of all signals of seismic phases that pass through the Earth's interior.

Regional High-Density Seismic Array

To obtain a better resolution for 3-D Earth models, we need seismic data from dense seismic arrays with a concentration of seismometers in a small region. The recent deployments of high density seismic arrays by a Japanese Agency, NIED (National Research Institute for Earth Science and Disaster Prevention), is an ideal example; the whole Japanese islands are covered with 1000 high-sensitivity seismometers at 20 km spacing. This is an unprecedented experiment of very high-density seismic network, providing an enormous amount of records. The data sets derived from the array are beginning to provide us with evidences of interesting phenomena inside the Earth. For example, we can "see" subtle movement of fluid in the lithosphere that is away from active volcanic fronts; i.e., it may be the first seismic evidence of the dehydration processes of subducting slabs in the mantle (Obara, 2002).

The deployment of such very high-density seismic arrays, starting with the tectonically active regions, will be of great importance in the progress of the solid Earth sciences in the coming decade.

Continent Probing with Portable Seismic Arrays

In seismically inactive areas, like the middle of most continents, it seems rather impractical to make a permanent installation of a high-density seismic array. However, such an array is essential to probe into the Earth. One of the most successful field experiments of deployment of portable seismic arrays at a continental scale has been conducted in Australia throughout the 1990's, (the SKIPPY Project – van der Hilst et al., 1994). Following its great success in producing high-resolution seismic models in the Australasian region, an extended experiment modeled on the SKIPPY philosophy, the USArray, is being conducted (e.g., Levander et al., 1999).

The USArray is now a part of the EarthScope project (United States), which includes SAFOD (San Andreas Fault Observatory at Depth), PBO (Plate Boundary Observatory) and InSAR (Interferometric Synthetic Aperture Radar) as well as the USArray to unveil the structure and evolution of the North American continent. The USArray proposes to cover the continental United States with a set of transportable networks of around 400 broad-band seismometers over ten years. The experiment will resemble a window blind pulled across the country with about 400 instruments deployed at a time, with a process of redeployment to the farther side after about 18 months in place. Thus, around 2000 stations in total are expected to cover the entire United States. The USArray will enable us to reveal the seismic structure of the North American Continent at an unprecedented resolution. Further developments of this type are to be expected for other continents.

Ocean Bottom Network

Seismic observations at the ocean bottom, which covers 70% of the Earth's surface, are critical to achieve uniform coverage of seismograph stations over the globe. Several projects to install seismometers at the ocean bottom are currently being undertaken, e.g., Ocean Hemisphere Project (University of Tokyo) and International Ocean Network. Continuous efforts to enhance such ocean bottom seismograph networks are required. There are efforts underway to develop disposable autonomous floating devices equipped with hydrophones in order to improve data collection in inaccessible regions without the human cost or ship time required to deploy and retrieve ocean bottom instruments (Simons et al., 2003). Moreover, up to now, there has been no means to measure deformation of the Earth's surface at the ocean bottom. Exploring seafloor deformation at GPS-level resolution will provide valuable information about dynamic processes of the Earth.

Interpretation and Analysis of Seismic Data

The enhancement of seismological interpretation and analysis will provide better data sets, such as better seismic event locations, better event catalogues, and more accurate and rich travel timetables of seismic phases, for further seismological researches. Also, physical parameters describing earthquake sources, such as radiated energy, seismic moment tensor, stress drop, rupture length, slip amount and slipping history, can be provided by seismological observation and interpretation on routine basis. Dissemination of such enhanced data sets will be an important contribution for a better understanding of earthquakes and physics of the Earth's interior.

2.3.2 Physics of the Earth's Interior: Structure, Composition and Dynamics

Seismology and related disciplines provide a means of looking into the inside of the Earth where dynamic activities on the geological time-scale are taking place. Our knowledge of the internal structure and dynamics of the Earth has advanced significantly in the last a few decades. This is mainly due to developments and advances in rapidly growing seismic networks, computer simulations of mantle convection and laboratory-based experiments on the physical properties of constituents of the Earth. This section focuses on future research topics that are expected and are necessary to be tackled in the coming decade for further understanding of the nature of the Earth's interior.

3-D Imaging of the Earth's Interior

Construction of 3-D Reference Earth Model: Extended seismograph networks have enabled us to delineate three-dimensional distribution of heterogeneity in the Earth. We now have plenty of seismic information that should be useful for constructing a three-dimensional Reference Earth Model, including large scale lateral heterogeneity in the Earth's mantle ($\sim$ 1000 km), which will be a substitute for conventional 1-D Reference Earth Models. In addition, creating a high-resolution (less than 100 km scale) reference crustal model will be crucial to enhance the resolution of regional tomography models. The information of an on-going project of creating a Reference Earth Model is available through a web site at UC San Diego (http://mahi.ucsd.edu/Gabi/rem.html).

Extending Seismic Tomography: The development of techniques for seismic tomography has brought a significant jump in our understanding of the Earth's interior, despite its relatively simple techniques mostly based on ray theory. Recently, many efforts have been made to obtain more precise and accurate images of the Earth's interior, taking account of much more complicated effects of seismic wave propagation, such as scattering and diffraction at finite frequency of seismic waves (Dahlen et al., 2000; Yoshizawa and Kennett, 2002). Such new approaches to seismic tomography require a significant amount of computation compared to the conventional ray theoretical methods, but will be applicable to large-scale inversion with the development of high-power computing facilities. Moreover, with powerful computers, a fully non-linear inversion, employing a global optimization method, can be applied to large-scale inverse problems. The evaluation of the ensemble of models derived from such direct non-linear inversions (e.g., Sambridge, 1999) will also be useful to validate 3-D Earth models.

Material Properties of the Earth's Interior

Physical Properties of Earth Materials and Seismic Models: What causes the lateral heterogeneity of seismic wave speed in the mantle? What is the role of water in the crust and mantle? Revealing rheological behaviour of the Earth materials is required to understand the nature of dynamic phenomena in the Earth, such as upwelling plumes, subducting plates, continental lithosphere, lithosphere-asthenosphere transition, etc. It is also important to address the relationship between wave speed anomaly, density anomaly and anelastic attenuation of the Earth's constituents working with *in-situ* laboratory experiments (e.g., Jackson, 2000).

Anisotropy and Dynamics of the Earth's Interior: There are a number of observations that suggest the existence of anisotropy in the crust and upper mantle (e.g., Park and Levin, 2002). The determination of the scale of anisotropy from seismic data is not simple because of trade-off between wave speed and anisotropy. Anisotropy of the Earth materials is likely to have a close relation with the current and historical dynamics of the Earth. Therefore, the study of the anisotropy in the Earth can be a clue to understand the convection processes of the mantle.

Dynamic Processes of the Earth's Interior

Integrated Earth Models: It is critical to understand large- and small-scale phenomena in the context of plate tectonics, convection processes, and Earth's rheology and material constitution. In this regard, studies of mantle convection and its effects on surface topography, plate tectonics, physical and chemical processes, and thermal sources in the mantle and core will be essential. Also, measurements of surface deformation, strain field and its variation by various geophysical approaches will provide useful constraints on the geodynamical models.

Seismic/Geophysical Discontinuities in the Earth: It is important to unveil the physical and chemical properties of seismic/geophysical discontinuities within the Earth, which play important roles in dynamic processes in the Earth. For example, Mantle Transition Zone (410 - 660 km depth), in which phase transformations in mantle minerals occur, plays a critical role in controlling the material circulation in the mantle. The existence of Ultra Low Velocity Zone (ULVZ) at the base of the mantle is suggested by many of recent researches. Further investigation of the role of CMB and D" and inner-core boundary are also important (see below).

Physics and Chemistry at the Core-Mantle Boundary (CMB): Of all the boundary layers in the Earth, the Core-Mantle Boundary and the D" layer are the least known, but a key region to understanding the thermal and dynamical systems of the Earth's mantle. It should be clarified what causes the strong heterogeneity in the D" layer, and what is a real figure of the upwelling plumes; a large-scale super plume or a cluster of small-scale plumes? Dense broad-band three-component seismic arrays will be a key to revealing the actual makeup of the D" (e.g., Kendall, 2000). Linking the high-resolution seismic evidences and numerical results of mantle dynamics will be essential for a better understanding of the bottom of the mantle.

Inner-core Rotation and Anisotropy: Is the inner-core rotating with respect to the mantle? The possibility of differential rotation of the inner core with respect to the mantle has been one of the topics in the centre of debate in the studies of Earth's deep interior. Some researches suggested that there is observable rotation of around 1 degree/year (Song and Richards, 1996), whilst others suggested negligible rotation rate (Laske and Masters, 1999). The elucidation of this issue will require long-term observations since the rotation rate seems to be too small to be adequately resolved without data sets over many years.

Recent studies also revealed that the inner-core is likely to have two distinct layers; the uppermost inner-core may be rather isotropic whereas the innermost inner-core may have anisotropic properties (Tromp, 2001). It should clarified how such a boundary between the innermost and outermost inner-core is formed. Revealing the state of the inner-core anisotropy is an important factor to clarify the formation and evolution of the solid core of the Earth. To obtain a better resolution of seismic models in the deep interior of the Earth, high-density seismic arrays will be of great help.

Fluid-Solid Interactions: What is the role of water and volatiles in the dynamics of the solid Earth? How does water affect plate tectonics, mantle convection and genesis of continents? In this respect, understanding the origin, distribution and the transportation of H_2O on the surface and the inside of the Earth is essential. The interactions of the solid Earth with ocean and atmosphere should also be investigated.

2.3.3 Physics of Earthquakes and Seismic Wave Field

Physics of Earthquake Sources

Generation of Earthquakes: Studies of the initiation and propagation of rupture processes at seismic sources are critical to better understand the physics of earthquakes. The governing equations for physical and chemical processes of seismic sources, as well as the Earth's mate-

rials, need to evolve from the microscopic scale to the macroscopic scale. Finding a bridge over such gaps based on physics is fundamentally important. Many unsolved problems in studies of fault dynamics, including stress state, dynamic rupture processes and the origin of spatio-temporal earthquake complexities, will need to be addressed in a common framework (e.g., Ben-Zion, 2001).

Role of Fluids in Fault Slip: In order to understand physical and chemical conditions that may lead to an immediate earthquake, interactions between fluid and rock will be of importance in the coming decade to understand the role of fluids in the facilitation of fault slip. Quantitative evaluation of effects of elastohydrodynamic lubrication of faults (Brodsky and Kanamori, 2001) may be a clue to understand the nature of rupture processes of faults.

Mechanisms of Deep Focus Earthquakes: There are also a number of puzzles in the mechanisms of the deep-focus earthquakes. Phase transition of the Earth's materials in the Mantle Transition Zone is likely to be one of the causes. Recent studies also suggested that dehydration processes in subducting slabs can be a cause of deep-focus earthquakes (Hacker et al., 2003). Further investigation of the physical and chemical processes, which may lead to the occurrence of deep-focus earthquakes should be pursued.

Modeling the Seismic Wave Field in 3-D Structures

Understanding the nature of seismic waves propagating in complex 3-D Earth models at both regional and global scales is indispensable for the progress of researches in seismic sources, Earth structure, and hazard mitigation. Recent high-power computing facilities, such as the Earth Simulator (Japan Marine Science and Technology Center), will enable us to simulate such numerical experiments, including high frequency seismic waves.

Recent advancement of PC cluster architecture with free operating systems, such as LINUX, provides an alternative high power parallel computing facility, which is extremely economical compared to costly supercomputers, and will be crucial for the development of computational seismology and geophysics (Komatitsch et al., 2002) (Fig. 2.2).

Scientific Basis for Comprehensive Test Ban Treaty (CTBT)

Improvements of the capability to identify the natural and man-made seismic events as well as the enhancement of global seismograph networks are essential to enforce the CTBT. A more accurate knowledge of 3-D Earth structure plays an important role in this regard. Global tomography models with the extended techniques as well as sophisticated

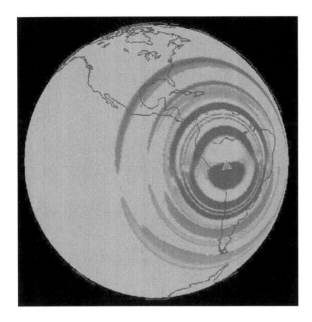

Figure 2.2: Snapshots of computer simulation of seismic waves generated by the June 9, 1994, Bolivia event. Full waveform modeling in a 3-D structure with rapidly-growing computing facilities will be a key to advancing the geosciences [courtesy of J. Tromp].

event location algorithms (e.g., Nicholson et al., 2002) will be of importance.

2.3.4 Hazard Mitigation and Prediction of Earthquakes

Risk Assessment and Hazard Mitigation

Taking precautions against catastrophic natural disasters is a critical issue for modern society. We should make continuous efforts to reduce the risks and damage to society. As yet, a systematic approach to accurate predictions for earthquakes is unavailable. In this regard, it is important to create hazard maps based on a quantitative assessment of the results of geophysical surveys in inhabited areas where there are high potential risks.

Simulating Strong Ground Motion

The most devastating damage inflicted by earthquakes is usually caused by strong ground motions near a seismic source. With improved knowledge about underground structures and the rupturing pattern of earthquake sources, we can simulate and assess how the ground will be shaken by incoming seismic waves (e.g., Furumura et al., 2002). Employing computer simulations of strong ground motion with an appropriate local structural model (e.g., Fig. 2.3), we can predict damage patterns and take some precautions against hazards that can

be caused by possible large earthquakes. Therefore, seismic exploration to reveal detailed seismic structures beneath metropolitan areas, in conjunction with earthquake engineering, will be a prerequisite for the mitigation of seismic hazards.

Quantitative Evaluation of Earthquake Precursors

Earthquake prediction has been debated for a long time. The opinions vary from optimism to extreme pessimism. The effort should be continued as long as humans inhabit the tectonically active areas of the Earth. There are some observations that seem to be related to earthquake precursors, but most of them are specific to particular regions and times, and seem not to be universal phenomena. The effort for finding such pre-earthquake phenomena is important for those regions where we expect high seismic risks. Quantitative evaluations of radio wave anomalies, electromagnetic phenomena, atmospheric precursors, chemical changes of groundwater, etc., in conjunction with physical and chemical properties of rocks, will be needed.

Early Warning System

In order to minimise damages of disastrous earthquakes to society, it is vital to provide information of early warning to the public immediately after the occurrence of the event. To achieve this, the quantitative assessment of the first motion of seismic signal is essential. This type of early warning system has already been introduced in some countries with high seismic risks, such as Japan, USA and Mexico. Dense permanent seismic networks are required to enable this system be useful in case of large earthquakes.

2.4 Recommendations for Seismology and Physics of Earth's Interior

The topics and strategies for future research in the framework of IASPEI discussed in the previous sections should be integrated to give a self-consistent dynamic model of the Earth, combining a variety of information sources from a wide range of scales in both time and space.

Linkage of Theory, Observation and Modeling:
Theoretical developments in seismology, such as wave propagation in 3-D Earth and the physics of earthquake sources, need to be merged with the rapidly growing number of observed data. The methods from the latest numerical modeling should play a role as a glue to link the latest theory and observations by helping interpretation of data and validation of the theory.

Simulating Scenario Disasters: The worst-case scenarios for earthquakes and tsunami disasters can be evaluated

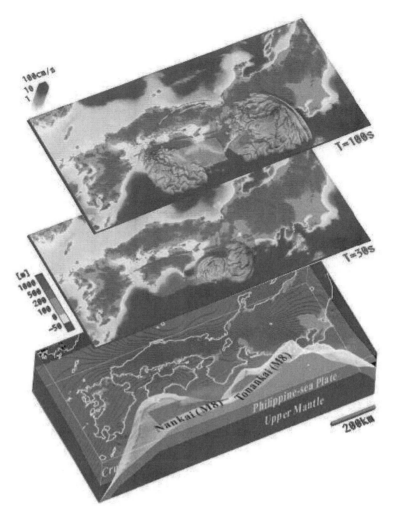

Figure 2.3: Snapshots of ground motion for Nankai "scenario" earthquake (Mw 8.0) in south-west Japan, derived from 3-D parallel simulation using the Earth Simulator. The bottom figure illustrates a 3-D model of the crust and upper mantle used in the wavefield modeling. Purple zones indicate rupture areas of the fault [courtesy of T. Furumura].

quantitatively by integrating a variety of state-of-the-art information and techniques, such as high-resolution seismic velocity models, the expected mechanism and magnitude of earthquake sources, sophisticated numerical methods, and high-power computing facilities (Fig 2.3). This will be helpful for the creation of seismic and tsunami hazard maps for regions with high risks of such disasters.

Gathering and Dissemination of Seismic Data: Covering the Earth uniformly with broad-band seismic stations is essential for the construction of high-resolution seismic models. International organizations, such as the Federation of Digital Broad-Band Seismograph Network (FDSN) and the Incorporated Research Institutions of Seismology (IRIS), have made efforts to facilitate circulation of seismic data by gathering data from regional seismic networks all over the globe. Still, the current coverage of seismic stations is not sufficient. It will be important to achieve a better global coverage of seismic networks by deploying new regional seismic arrays, including ocean bottom instrumentation, as well as by gathering data from existing regional networks, which are yet to join the FDSN.

Fostering International Schools for Young Researchers: Promoting educational programs for students and young researchers is important to fostering international collaboration amongst young generations. Financial support for young researchers, especially those from developing countries, should be broadly available.

International training programs for researchers from developing countries will also be important to educate and train scientists, technicians and science teachers in those regions. There is a successful example of an international training project, which has been offered by the Building Research Institute of Japan for more than 40 years, for educating professionals and technical experts in seismology and earthquake engineering from developing countries.

Promoting International Projects: It will be one of the important missions of IASPEI to foster international collaborative observations and research projects. In particular, cooperation between developing and developed countries will contribute to both the development of global seismology and the development of regional economy and society in developing countries.

Efficient Use of Internet Resources: The internet provides us with enormous possibilities for easy and economical communications with colleagues all over the world. With the advent and spread of the Giga-bit networks, it will be possible to hold face-to-face international conferences, schools, workshops, and a variety of meetings through the internet. Free access to geoscientific information through the internet would be of great help in the advancement of geosciences.

2.5 Seismological Researches and Societal Benefit

Hazard Maps: It is essential to create hazard maps to evaluate the seismic risk in local areas, based on geological and geophysical investigations of active faults and numerical simulations of strong ground motion in a local seismic structure. Basic researches on crustal deformations and seismicity pattern analysis are also important.

Communication with Decision Makers: In case of catastrophic disasters, scientific information must be smoothly transmitted to decision makers to take prompt actions against hazards. Geoscientists should keep in mind that scientific information on natural disasters should be provided in layman's words to avoid miscommunication.

Information Technology: It will be useful to create a system using information technology as a communication tool in case of emergency. Outer space monitoring with satellites for earthquakes and disasters may be helpful in immediate transmission of rescue information. It is also important to utilize rapid and reliable earthquake ground-motion information for controlling large structures and complex systems to minimize the impact of large earthquakes on modern society.

Building Code and Quakeproof Building: In seismically active areas, the legislation of appropriate building codes is essential. Engineering solutions to strengthen buildings against strong ground motions at a reasonable cost will be of significant help in reducing damages in developing countries.

Insurance Policy: Disastrous earthquakes in urban area cause not only large numbers of casualties, but also huge economic losses. Insurance policies for the case of catastrophe should be constructed based on the latest scientific information.

Personal Seismic Monitoring System: It will be useful to disseminate personal sensors and recoding systems that can be linked to personal computers, internet, TV, and/or mobile telephones, as a complement of professional observation facilities. Such personal facilities can play a role as a component of education of the public and outreach for the reduction of natural disasters.

2.6 Opportunities for Interdisciplinary Researches

Linking Surface Tectonics and Mantle Dynamics (IAG, IAVCEI): The deformation of surface topography reflects the dynamic processes that are going on in the Earth. Continuous geodetic observations of the surface deformation provide not only useful information for understanding the dynamics of the Earth's interior, but also indispensable information on the occurrence of earthquakes and volcanic eruptions.

Volcano Seismology (IAVCEI, IAG): Most volcanic activities are accompanied by seismic activities, and thus seismic observation in volcanic regions is inevitable. Revealing the processes of interaction between earthquakes and volcanic activities would be a clue to better understanding the triggering effects between eruptions and earthquakes. Furthermore, seismology can provide useful tools for the development of physical volcanology. To obtain a detailed image of seismic structure of volcanoes, enhanced seismic observations in volcanic areas as well as the advancement of theories and techniques of seismic wave propagation and modeling in strongly heterogeneous media will be essential.

Generation and Propagation of Tsunami (IAPSO): Earthquakes in oceanic regions sometimes generate huge tidal waves, tsunamis, which bring catastrophic damages in coastal areas. The scientific investigation of its generation and propagation is required to create a practical warning system to mitigate tsunami disasters.

Seismology for the Earth and the Sun (IAGA): The studies of free oscillations of the Earth and the Sun share some common theoretical ground (e.g., Dahlen & Tromp, 1998). Further developments of seismology would complement helioseismology, and vice versa.

Coupling Between Atmosphere, Ocean and Solid Earth (IAMAS, IAPSO, IAG): How do the solid Earth, atmosphere and ocean interact with each other? How do the atmosphere and ocean shake the solid Earth? The development of a systematic Earth model integrating a variety of Earth processes, such as global warming, growth and retrogression of polar ice caps, plate tectonics, continental evolution, etc, which fully incorporate the physics and chemistry of the processes into their structure, considering the significant differences in time-scale, will be a common ultimate goal of geosciences.

Water Cycle in the Earth (IAHS, IAPSO, IAVCEI): How is water brought down into the deep Earth? What is the role of the water in the Earth? Can water be a catalyst for earthquakes? The process of transportation of groundwater and its implications to the precursory anomalies of earthquakes will need to be evaluated quantitatively.

2.7 Conclusions

There are a number of areas of challenging research related to IASPEI to extend our scientific knowledge in the future. Enhanced information about the Earth's interior is important for a better understanding of the present status of the Earth as well as its historical evolution as a planet. Unveiling the nature of earthquakes and their generation, incorporating a variety of information of relevant disciplines, will be essential to meet the societal needs, i.e., earthquake prediction and mitigation of their effects. The rapid advancement of modern technology will make it possible to open new fields of interdisciplinary sciences. The first decade of the 21st century will be the exciting years for the younger generation to achieve a further understanding of the Earth's interior as well as a dynamic system of the whole Earth.

2.8 Acknowledgment

Some of the ideas outlined in this chapter have arisen from answers to a survey conducted in 2002. The author would like to thank all of those who kindly shared their opinions; in particular, B. Chouet, E. R. Engdahl, H. Kanamori and K. Suyehiro for their comprehensive comments on important issues of the IASPEI. I would also like to thank E. R. Engdahl, B. L. N. Kennett, P. Suhadolc, Z. Wu, F. Simons and T. Hara for their reviews which helped to improving the report.

Bibliography

Ben-Zion, Y, Dynamic ruptures in recent models of earthquake faults *J. Mech. Phys. Solids*, **49**, 2209-2244, 2001.

Brodsky, E. E. and H. Kanamori, Elastohydrodynamic lubrication of faults, *J. Geophys. Res.*, **106**, 16357-16374, 2001

Dahlen, F. A., S. -H. Hung and G. Nolet, Fréchet kernels for finite-frequency travel times - I. Theory, *Geophys. J. Int.*, **141**, 157-174, 2000.

Dahlen, F. A. and J. Tromp, *Theoretical Global Seismology*, Princeton University Press, Princeton, 1998.

Furumura, T., K. Koketsu,.and K.-L. Wen, PSM/FDM hybrid simulation of strong ground motion for the 1999 Chi-Chi, Taiwan earthquake, *Pure Appl. Geophys.*, **159**, 2133-2146, 2002.

Hacker B.R., S. M. Peacock, G. A. Abers and S. D. Holloway, Subduction factory - 2. Are intermediate-depth earthquakes in subducting slabs linked to metamorphic dehydration reactions? *J. Geophys. Res.*, **108**, ESE 11, 10.1029/2001JB001129, 2003.

Jackson, I. Laboratory measurement of seismic wave dispersion and attenuation: Recent Progress, in *Earth's Deep Interior: Mineral Physics and Tomography From the Atomic to the Global Scale*, pp. 265-289, ed. Karato, S., Forte, A. M., Liebermann, R. C., Masters, G. and Stixrude, L., American Geophys. Union, 2000.

Kendall, J.-M., Seismic anisotropy in the Boundary Layers of the Mantle, in Earth's Deep Interior: Mineral Physics and Tomography From the Atomic to the Global Scale, pp. 133-159, ed. Karato, S., Forte, A. M., Liebermann, R. C., Masters, G. and Stixrude, L., American Geophys. Union, 2000.

Kobayashi N. and K. Nishida, Continuous excitation of planetary free oscillations by atmospheric disturbances, *Nature*, **395**, 357-360, 1998.

Komatitsch, D., J. Ritsema and J. Tromp, The spectral-element method, Beowulf computing and three-dimensional seismology, *Science*, **298**, 1737-1742, 2002.

Laske, G. and G. Masters, Limits on differential rotation of the inner core from an analysis of the Earth's free oscillations, *Nature*, **402**, 66-69, 1999.

Lee, W. H. K., H. Kanamori, P. C. Jennings and C. Kisslinger, International Handbook of Earthquake and Engineering Seismology, Part A, Amsterdam; Tokyo, Academic Press, 2002.

Lee, W. H. K., H. Kanamori, P. C. Jennings and C. Kisslinger, International Handbook of Earthquake and Engineering Seismology, Part B, Amsterdam; Tokyo, Academic Press, Academic Press, 2003.

Levander, A., E. G. Humphreys, G. Ekstrom, A. S. Meltzer and P. M. Shearer, Proposed project would give unprecedented look under north America, *EOS Trans. AGU*, **80**, 250-251.

Nicholson, T., M. Sambridge and O. Gudmundsson, Hypocenter location by pattern recognition, *J. Geophys. Res.*, **107**, ESE 5, 10.1029/2000JB000035, 2002.

Obara, K., Nonvolcanic deep tremor associated with subduction in southwest Japan, *Science*, **296**, 1679-1681, 2002.

Park, J. and V. Levin, Geophysics - Seismic anisotropy: Tracing plate dynamics in the mantle, *Science*, **296**, 485-489, 2002.

Sambridge, M., Geophysical Inversion with a Neighbourhood Algorithm -II. appraising the ensemble, *Geophys. J. Int.*, **138**, 727-746, 1999.

Simons, F. J., G. Nolet and D. Bohnenstiehl, Scrap the cable: identification and discrimination of seismic phases by autonomous floats, *Geophys. Res. Abstracts*, **5**, 08035, European Geophysical Society, 2003.

Song X. and P.G. Richards, Seismological evidence for differential rotation of the Earth's inner core, *Nature*, **382**, 221-224, 1996.

Suda N, K. Nawa, Y. Fukao, Earth's background free oscillations, *Science*, **279**, 2089-2091, 1998.

Tromp, J., Inner-core anisotropy and rotation, *Ann. Rev. Earth Planet. Sci.*, **29**, 47-69, 2001.

van der Hilst, R., B. Kennett, D. Christie and J. Grant, Project Skippy explores the lithosphere and mantle beneath Australia, *EOS Trans. AGU*, **75**, 177-182, 1994.

Yoshizawa K. and B.L.N. Kennett, Determination of the influence zone for surface wave paths, *Geophys. J. Int.*, **149**, 440-453, 2002.

Zhao, D., H. Kanamori, H. Negishi, Tomography of the source area of the 1995 Kobe earthquake: Evidence for fluids at the hypocenter?, *Science*, **274**, 1891-1894, 1996.

Chapter 3

International Association of Volcanology and Chemistry of the Earth's Interior

E. E. Brodsky

The International Association of Volcanology and Chemistry of the Earth's Interior (IAVCEI) promotes both research in volcanology and efforts to mitigate volcanic disasters. IAVCEI also promotes research in the closely related fields of igneous petrology and geochemistry. As the major scientific problem of volcanology is also one of societal import, the work and progress of the field is closely related to the individual situations in local communities. The value of an international organization like IAVCEI is to help integrate knowledge across boundaries, both political and intellectual. By introducing this broad perspective, IAVCEI helps volcanologists turn their discussions from fire-fighting to strategies for pushing forward a scientific frontier.

3.1 Long-term goals

We would like to know accurately what are the processes leading to an eruption, and how do these processes relate to observable precursors such as volcanic tremor, earthquakes, gas discharge events and deformation. Why do volcanoes erupt; i.e., why does the magma not remain as an intrusion at depth? We would like to be able to use this knowledge to accurately forecast not only when an eruption will occur, but also the size and type of the eruption. We should also be able to distinguish an impending eruption from mere unrest.

This wish list may seem difficult to achieve, but we are not as far from it as some might imagine. Eruptions at well-monitored volcanoes seldom occur without warning. In broad outlines, we know that volcanoes erupt because the overpressure of the volatiles exceeds the strength of the confining material. However, the quantitative details behind this scenario are still elusive. The type of the eruption at a given volcano, including its potential explosiveness, is usually forecast based on the previous history of the volcanic system with varying degrees of accuracy. Distinguishing true impending eruptions from dike intrusions or merely a seismic swarm is still a difficult matter that is often decided based on expert opinion and experience rather than quantifiable scientific predictions. The physical process behind certain precursors, like inflation, may be well-known, but others, such as tremor, remain enigmatic. Even the relatively well-understood precursors are associated with only isolated pieces of the magmatic system rather than a comprehensive view of the magma-volatile-wallrock evolution.

3.2 Short-Term Priorities

In order to reach the long-term goal of quantifiable predictions based on physical interpretations of precursors, we need to concentrate on unraveling specific volcanic processes. The following section highlights a few major questions in volcanology that could plausibly be answered by focused efforts in the next 10–20 years. The questions are organized by process rather than methodology or discipline because answering most problems will require a combination of techniques.

Strategies for achievement follow each set of questions. A limited number of example studies are cited to illustrate and clarify the strategies suggested. The citations here are by no means a complete survey of all work done in the field nor is this discussion an exhaustive list of interesting problems in volcanology. The purpose of this report is to focus on opportunities for rapid and important advancement of the basic science.

Many of the important problems do not require state-of-the-art technology or large-scale research investments. Several of the suggested research strategies are achievable by single investigators in regions without major technological infrastructure. Scientists in developing countries will likely continue to make fundamental contributions to the field.

3.2.1 Initiation

What begins an eruption? Is it the failure of the edifice or the pressurization of the chamber due to entirely internal properties? What is the role of external sources of volatiles such as groundwater? Is the intrusion of new magma or an episode of mixing necessary to start a new eruptive episode?

Progress could be made here by detailed studies of instrumental records of the first few minutes of an eruption, petrological studies including kinetic constraints on magma rise rates [*Rutherford and Hill*, 1993; *Geschwind and Rutherford*, 1995], and comparison of precursory activity across many examples of historical eruptions [*Newhall and Dzurisin*, 1988; *Benoit and McNutt*, 1996].

3.2.2 Explosive/effusive transition

What combinations of volatile abundances and magma chemistries lead to explosive rather than effusive eruptions? Are there any observable indicators of the state prior to eruption for a volcano that has a history of both types of behavior?

Promising leads on the answer to the first question have come from laboratory experiments on natural materials and analog studies [*Martel et al.*, 2001; *Ichihara et al.*, 2002]. Experimental petrological studies of silicate rheology as a function of composition, volatile content, crystallinity, temperature and pressure connect the analog studies to the natural systems. The observational question will likely be answered by well-integrated modeling and observational studies of very active volcanic systems like the ongoing Soufriere Hills, Montserrat eruption that regularly produces both types of activity [*Druitt and Kokelaar*, 2002].

3.2.3 Eruptive Plumes

What are the expected plume heights and dispersal patterns for realistic eruptions?

Much theoretical progress has been made on plume heights from idealized eruptions [*Wilson et al.*, 1980; *Sparks and Wilson*, 1976; *Valentine and Wohletz*, 1989; *Neri and Dobran*, 1994; *Woods*, 1995], yet there still are comparatively few studies comparing the predictions to field data. Field data now encompasses satellite images as well as more conventional maps of deposits [*Krotkov et al.*, 1999]. Comparisons with the data will likely require incorporating realistic multiphase properties, large depressurizations, 3-D effects and compressibility into a new generation of numerical models.

3.2.4 Climatic effects of eruptions

Can we predict the global pattern and amplitude of temperature decrease from a given volcanic eruption? How regionally variable are these effects? What is the long-term role of volcanic volatiles in the atmospheric mass budget?

The raw data of the climate and volcanological record will come from a combination of geological studies, tree rings and especially ice cores. Numerical climate modeling with Global Climate Models (GCM) informed by an accurate understanding of the volcanological inputs and uncertainties can address these questions. Gas monitoring data including isotopic compositions are necessary inputs to volatile mass budget considerations [*Hilton et al.*, 2002].

3.2.5 Flow Emplacement

What determines the runout distance of pyroclastic flows? Of lava flows? Of lahars?

Geological deposits record the emplacement of flows and so can be interpreted to determine the processes ending movement [*Branney and Kokelaar*, 1992]. Laboratory analog experiments have had considerable success in quantifying the morphology of lava flows in terms of dimensionless parameters; similar approaches can be taken to runout length [*Fink and Griffiths*, 1998]. Laboratory studies of silicate melt rheology again help relate the analog materials to the real systems. Images of flow on other planets can extend our parameter ranges to test theories [*Zimbelman*, 1998].

3.2.6 Interactions with Tectonics

When do earthquakes trigger eruptions? When do eruptions trigger earthquakes? How?

Triggering problems can be approached with statistical studies of the historic record [*Linde and Sacks*, 1998] as well as examining monitoring records (thermal, gas, hydrothermal, deformation and seismic data) for evidence of coupling [*Johnson et al.*, 2000].

3.2.7 Magma Storage

Are magma storage areas chambers or plexuses? Is the storage body created by a single emplacement event or repeated injections? What is the relationship, if any, between storage body size and eruptive volume? Do magma bodies convect and if so, for what parameter ranges? What is the volatile content in the storage areas?

Geological studies of either erupted ignimbrites and emplaced plutons give the most direct data on magma chamber heterogeneity [*Hildreth and Fierstein*, 2000]. Laboratory and theoretical studies could connect microstructure and zonation to specific parameters. Seismic and other geophysical methods provide direct constraints on the geometry of active bodies [*Auger et al.*, 2001]. Ultimately, the petrological story should be connected to the physical story.

3.2.8 Conduit Transport

At what stage of an eruption is the conduit geometry deter-

mined? When is magma transport by percolation rather than conduit flow? What role does thermal buoyancy play? Does the collapse of conduits end eruptions? Where and how does degassing take place in a conduit?

Coupled solid-fluid numerical models will likely be an important part of answering these questions. Magmatic rheology is once again a crucial input to the models. Geological studies of wallrock entrainment and abundance will pose fundamental constraints for the models. Gas studies constrain the degassing history.

3.3 Recommendations for Volcanology

In general, attacking these problems will require combining multiple techniques and comparing multiple eruptions across different volcanoes. The above discussion suggests two general categories for improvement: (1) synthesis of observations and modeling and (2) instrumentation.

3.3.1 Observations and Modeling

Although vast technological progress has been made in observing volcanoes and major conceptual and numerical leaps have been made in modeling them, there is still a fundamental lack of connection between these two major sides of volcanology. The inputs to the models are generally not the observables in the field. Theorists would like to know magma chamber volatile content and viscosity while field scientists report number of earthquakes and gas discharge. Closing this gap is going to require efforts from both sides. The modelers need to develop models that include degassing and deformation. A model that is not connected to data is useless. The field scientists are going to have to invert their results for properties in the magma. Numbers that are not connected to a real process are meaningless. New initiatives like the WOVOdat project discussed below could broadly disseminate the raw observational data. This distribution will be an important tool for making progress.

The study of volcanic seismicity is a special case of the interface between observation and modeling. Many groups consider volcanic seismic signals to be the most reliable precursors to an eruption, and therefore, their physical interpretation is a priority in the field. However, the interpretation of these precursors is still so preliminary that it was not possible to work volcano seismicity into the process-oriented short-term priorities of the previous section. Is tremor giving information about the initiation process, conduit transport or some other physics? Volcano seismologists today are like the early petrologists who first organized a taxonomy of observations and only later developed an understanding of their genesis. Just as the early rock classifications that placed limestone and diabase together caused much confusion, it is almost inevitable that the currently used empirical seismicity classification is misleading in some respects.

Overcoming these barriers will require combining data and models by asking questions like: Where are the earthquakes relative to the magma? Is their location controlled by the stress field or local structures at the hypocenters? What controls the frequency of volcanic seismicity?

High-resolution earthquake relocations provide powerful insights into the question of location [*Gillard et al.*, 1996]. Well-integrated studies of these locations and the theoretical predictions of coupled fluid-solid models can explain why the earthquakes are located in the observed places [*Rubin et al.*, 1998]. Connecting the frequency of volcanic tremor and long-period events to physical processes may benefit from comparing the volcanic timeseries with other observables such as gas, chemistry or deformation. Forward models that predict multiple observables based on coupled fluid and solid mechanics will likely be a necessary step in solving this problem and others like it.

3.3.2 Instrumentation

As a natural science, volcanology is propelled forward by observational data. Quantitative instrumental data connects observations to processes and ultimately to predictive models. The last decade has seen major improvements in instrumentation, including the widespread use of InSAR (Figure 3.1), GPS monitoring and broadband seismometers. Increasingly, active eruptions are being monitored and forecast by primarily instrumental means. The education of the next generation of volcanologists must take this fact into account. In addition, further technological developments are needed to address the following issues.

The narrowing geodesy/seismology bandwidth gap

The classical instrumentation of volcanic geodesy and seismology has converged. Broadband seismometers can measure velocity and displacement from 50 Hz to hundreds of seconds. Real-time kinematic GPS systems monitor displacement changes over seconds or months and years. Gravimetry has historically been used to measure changes over years, but now modern digitizers allow data to be logged over seconds. These overlapping bandwidths have some distinct advantages. We can now completely cover the spectrum and can reasonably compare data from diverse tools. However, there are also some subtle problems. Have certain techniques become redundant? The interpretation of a classical technique outside its classical bandwidth can be problematic and plagued by response issues as instrumentation is pushed to its limits. It now seems within reach to design a single motion sensor over all relevant wavelengths. Such a leap could

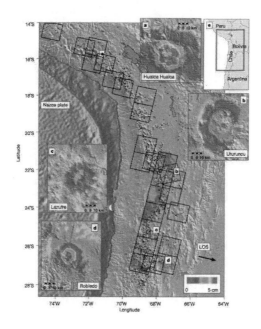

Figure 3.1: Example of InSAR deformation data on a shaded relief topographic map of the central Andes. Black squares show the radar frames used in this remote-sensing study. Black triangles show the 1113 potential volcanic edifices. The red line in the ocean is the location of the trench. Color shows InSAR interferograms indicating active deformation. Each color cycle corresponds to 5 cm of deformation in the radar line-of-sight (LOS) direction. Inset maps provide detailed views of the centers of volcanic deformation: a., Hualca Hualca, Peru, time span 6/92–4/96 (3.9 yr), b., Uturuncu, Bolivia, time span 5/96–12/00 (4.6 yr), c., Lazufre, time span 5/96–12/00 (4.6 yr), all of which are inflating, and d., Robledo, time span 5/96–10/00 (4.4 yr), in NW Argentina which is deflating (From *Pritchard and Simons* [2002]).

greatly reduce the cost of volcanic monitoring while increasing the ease of interpretation of datasets.

Remote sensing & robotics

Fieldwork on active volcanoes is inherently dangerous. Increasingly, technology is able to replace hand-sampling and in-person monitoring. Satellite remote sensing has made great strides in measuring deformation, thermal anomalies, ash dispersion and gas discharge. Satellites provide excellent spatial coverage, but they do not resolve rapid time-dependent processes or trace chemical compositions. Although we should certainly attempt to expand satellite data and its availability, we will still need locally deployed instruments to measure such staples of the field as seismic waves and fumarole chemistry. There have been some efforts to use robotics to deploy such nearfield instruments, no system has yet been completely successful. A renewed push on this front is necessary in order to assure high-temporal resolution intracrater monitoring.

Gas monitoring

Theoretical studies indicate volatile abundance is a major parameter in determining eruptive style. Observational constraints on *in situ* volatile concentrations coupled to a model of degassing and partitioning can provide a useful starting point for predictive modeling of eruptive dynamics. SO_2 concentrations have been measured for decades, but it is only recently that other components of the plume have begun to be measured. Instrumentation, such as FTIR that measures multiple volatiles, should be developed and deployed routinely.

Real-time integrated datastreams

Modern telemetry makes it possible to collect large quantities of data in real-time. Some techniques, like seismology, have already been able to capitalize on this ability, while others, like gas monitoring, are still developing the technology. Real-time simultaneous display of multiple observables has obvious advantages for hazard management. It would also benefit basic research by both facilitating the integration of data types and providing the framework for ready comparison of models and data. In particular, if all the relevant observables were acquired as digital datastreams, whole new methods of timeseries analysis would become possible. Theoretical predictions could be tested in a forward sense. Formal data assimilation techniques could be used to update and validate physical models. Even more speculative methods, like timeseries continuation, might someday become practical prediction tools and windows into physical processes.

3.4 Recommendations for IAVCEI and IUGG

3.4.1 Travel for Observatory Scientists

IAVCEI should raise funds specifically designated to bring volcanic observatory scientists from the developing world to major conferences such as AGU, EUG and IAVCEI. This central observatory travel fund should be a priority as it would address several issues. The fund could help observatory scientists present their data themselves. The fund could also facilitate modeling-observation interactions by helping observers appreciate state-of-the-art scientific questions and allow modelers direct access to observers' experiences. It could also help observatory scientists discuss logistical and procedural matters across international boundaries that could improve overall data quality and provide a uniformity of observations that is required if we are to have true cross-volcano comparative studies.

3.4.2 WOVOdat

Interdisciplinary research could be facilitated by disseminating a variety of data in a uniform format to be accessible to anyone without any special connections. WOVOdat (http://www.wovo.org/wovodat.htm) is a project to do just this and deserves IUGG's full and active support. The project would also be facilitated if the observatory scientists were better connected to the research process as suggested in the above discussion of a the central observatory travel fund.

3.4.3 Communicating with Society

Scientific progress will not mitigate hazards unless results are communicated effectively to local decision makers during a crisis. Over the past 20 years, much attention has been paid to the interface between scientists and politicians with significant practical and scholarly research on hazard management. Still, sometimes turf wars and scientists inexperienced in crisis management lead to difficult and dangerous situations. As a new generation of scientists emerges, we all need to be constantly reminded of the seriousness of our task and retrained on appropriate ways to interface with local communities. Crisis teams must effectively combine the local knowledge of resident scientists and observers with the constantly evolving skills of foreign specialists. Technical results must be made accessible without sacrificing factual accuracy and completeness. In order to facilitate this process, IAVCEI should continue to serve as a central source of guidelines and training on crisis management.

3.5 Opportunities for Interdisciplinary Research

Volcanoes literally explode the boundaries between the spheres of geoscience. Opportunities for interaction exist between IAVCEI and each of the other associations. Possibilities include:

- IAMAS: Exploration of the effects of volcanism on climate (Section 3.2.4).

- IASPEI Earthquake-volcano interactions (Section 3.2.6). Development of robust wave propagation modeling tools suitable for the highly heterogeneous paths of volcanoes. Characterization of explosive and other non-double couple seismic sources. Delineation of geometry of magma bodies (Section 3.2.7).

- IAG: Geodetic instrumentation at high time-resolution (Section 3.3.2).

- IAGA: Numerical and theoretical tools for modeling large-scale compressible jets (Section 3.2.3).

- IAPSO: Mid-ocean ridge volcanism and earthquake interactions (Section 3.2.6). The role of submarine volcanism in the biological and physical environment of ridges.

- IAHS: Groundwater as a trigger for explosive eruptions (Section 3.2.1). Using water well data to measure strain, temperature and chemical changes in the coupled hydrothermal-volcanic system.

3.6 Conclusions

Recent instrumental advances and the broad dissemination of observatory data are bringing volcanology into a new era. IAVCEI and the volcanology community can take advantage of the situation by targeting our resources to answer the questions most likely to generate a quantitative, portable model of why volcanoes erupt.

3.7 Acknowledgments

The author is grateful for responses to a survey in September 2002, especially C. Newhall, who contributed a complete report on the intersection of modeling and observations. Several reviewers including K. Cashman, T. Fischer, M. Manga, M. Mangan, S. Prejean, M. Pritchard and R.S.J. Sparks contributed to improving the report.

Bibliography

Auger, E., P. Gasparini, J. Virieux, and A. Zollo, Seismic evidence of an extended magmatic sill under Mt. Vesuvius, *Science*, *294*, 1510–1512, 2001.

Benoit, J.R., and S.R. McNutt, Global volcanic earthquake swarm database 1979–1989, *USGS Open-File Report*, *96-69*, 1996.

Branney, M.J., and B.P. Kokelaar, A reappraisal of ignimbrite emplacement: changes from particulate to non-particulate flow during progressive aggradation of high-grade ignimbrite, *Bull. Volcanol.*, *54*, 504–520, 1992.

Druitt, T.H., and B.P. Kokelaar, *The eruption of Soufrire Hills Volcano, Montserrat, from 1995 to 1999*, Geological Society, London, 2002.

Fink, J.H., and R.W. Griffiths, Morphology, eruption rates, and rheology of lava domes: Insights from laboratory models, *J. Geophys. Res.*, *103*, 527–545, 1998.

Geschwind, C.H., and M.J. Rutherford, Crystallization of microlites during magma ascent: The fluid mechanics of 1980–86 eruptions at Mount St. Helens, *Bull. Volc.*, *57*, 356–370, 1995.

Gillard, D., A. M. Rubin, and P. Okubo, Highly concentrated seismicity caused by deformation of Kilauea's deep magma system, *Nature*, *384*(6607), 343–346, 1996.

Hildreth, W., and J. Fierstein, Katmai volcanic cluster and the great eruption of 1912, *Geol. Soc. Am. Bull.*, *112*, 1594–1620, 2000.

Hilton, D. R., T. P. Fischer, and B. Marty, *Noble gases in geochemistry and cosmochemistry*, chapter Nobles Gasses and Volatile Recyclying at Subduction Zones, pages 319–362, Mineral. Soc. of Amer., 2002.

Ichihara, M., D. Rittel, and B. Sturtevant, Fragmentation of a porous viscoelastic materials: implication to magma fragmentation, *J. Geophys. Res.*, *106*(B10), 2229, 2002.

Johnson, H.P., M Hutnak, R.P. Dziak, C.G. Fox, I. Urcuyo, J.P. Cowen, J. Nabelek, and C. Fisher, Earthquake-induced changes in a hydrothermal system on the Juan de Fuca mid-ocean ridge, *Nature*, *407*, 174–177, 2000.

Krotkov, N. A., O. Torres, C. Seftor, A. J. Krueger, A. Kostinski, W. I. Rose, G. J. S. Bluth, D. Schneider, and S. J. Schaefer, Comparison of TOMS and AVHRR volcanic ash retrievals from the August 1992 eruption of Mt. Spurr, *Geophys. Res. Lett.*, *26*(4), 455–458, 1999.

Linde, A.T., and I.S. Sacks, Triggering of volcanic eruptions, *Nature*, *395*, 888–890, 1998.

Martel, C., D.B. Dingwell, O. Spieler, M. Pichavant, and M. Wilke, Experimental fragmentation of crystal- and vesicle-bearing silicic melts, *Bull. Volcanol.*, *63*, 398–405, 2001.

Neri, A., and F. Dobran, Influence of eruption parameters on the thermofluid dynamics of collapsing volcanic columns, *J. Geophys. Res.*, *99*, 11,833–11,857, 1994.

Newhall, C.G., and D. Dzurisin, Historical unrest at large calderas of the world, *U.S. Geol. Surv. Bull.*, *1855*, 1988.

Pritchard, M.E., and M. Simons, A satellite geodetic survey of large-scale deformation of volcanic centres in the central Andes, *Nature*, *418*, 167–171, 2002.

Rubin, A. M., D. Gillard, and J. L. Got, A reinterpretation of seismicity associated with the January 1983 dike intrusion at Kilauea Volcano, Hawaii, *J. Geophys. Res.-Solid Earth*, *103*(B5), 10003–10015, 1998.

Rutherford, M.J., and P.M. Hill, Magma ascent rates from amphibole breakdown: Experiments and the 1980–1986 Mount St. Helens eruptions, *J. Geophys. Res.*, *98*, 11949–11959, 1993.

Sparks, R. S. J., and L. Wilson, A model for the formation of ignimbrite by gravitational column collapse, *J. Geol. Soc. Lond.*, *132*, 441–451, 1976.

Valentine, Greg A., and Kenneth H. Wohletz, Numerical models of plinian eruption columns and pyroclastic flows, *J. Geophys. Res.*, *94*, 1867–1887, 1989.

Wilson, Lionel, R. S. J. Sparks, and George P. L. Walker, Explosive volcanic eruptions; IV, the control of magma properties and conduit geometry on eruption column behaviour, *Geophys. J. R. Astr. Soc.*, *63*, 117–148, 1980.

Woods, A.W., The dynamics of explosive volcanic eruptions, *Rev. Geophys.*, *33*, 495–530, 1995.

Zimbelman, J. R., Emplacement of long lava flows on planetary surfaces, *J. Geophys. Res.*, *103*, 27,505–27,516, 1998.

Chapter 4

International Association of the Physical Sciences of the Oceans

C. Simionato

4.1 Introduction

The aim of physical oceanography is to understand, model and predict ocean processes by combining observations with mathematics and fluid mechanics. Physical oceanography is concerned with how water moves and mixes in the ocean and how water transports and distributes heat and dissolved and/or in suspension chemicals, nutrients, plankton, sediment, and pollutants. This discipline does not only include the study of the large oceans and shelf seas but also of estuaries, lakes and large bodies of water on other planets and moons.

Physical oceanography is an extremely challenging science. One of its central difficulties, in common with other geosciences, is the lack of a laboratory to experiment in. Physical oceanographers must rely, therefore, on direct observations and interpretations of these data with fluid mechanics, applied mathematics, powerful computers and modern descriptive tools. Another great challenge is the range of space and time scales that must be measured and modeled by any successful effort to understand the fluid behavior. Every scale, from the seconds to hours of the waves and tides that modify the shoreline to the centuries and even millennia of the ocean thermohaline circulation response scale, going through a number of intermediate scales, must be considered.

Comprehensive observation of these processes demands a difficult and expensive combination of in situ and remote measurements for very long periods of time and large distances, which just started in relatively recent years. Other not minor tasks include the need for an international cooperation effort to observe and understand the system, given that the ocean knows no national boundaries but behaves as a whole, and the need for an interdisciplinary approach in order to give answers to many important environmentally relevant questions.

These needs make progress in physical oceanography to be intrinsically tied to progress in technology. Recent advances in observational tools, especially in remote sensing, together with the development of new and powerful computer based tools are producing an increasingly global and complete picture of the three-dimensional ocean circulation and great advances in our physical understanding of the oceans. The field is beginning to assemble the elements to reach the ultimate scientific goal of attaining a comprehensive understanding of the physics of the ocean.

The International Association of the Physical Sciences of the Oceans (IAPSO) is one of the seven associations that comprise the International Union of Geodesy and Geophysics (IUGG), founded in 1919. IAPSO has as its prime goal "promoting the study of scientific problems relating to the oceans and the interactions taking places at the sea floor, coastal, and atmospheric boundaries insofar as such research is conducted by the use of mathematics, physics, and chemistry." One of the main objectives of IAPSO is to organize, sponsor, and co-sponsor formal and informal international forums, permitting ready means of communication amongst ocean scientists throughout the world. In this sense, the IUGG general assembly in Sapporo, Japan, provides an excellent context for reflection and interdisciplinary discussions about our vision of how geosciences should develop during the next decades. This report intends to provide a view of the main directions in which physical oceanography research could progress during that period and to serve as a starting point for those discussions.

The results of previous workshops and surveys, together with a survey and revision process among several scientists around the world were very valuable in the preparation of this report.

4.2 A view of the future of research in Physical Oceanography

Even though basic research, independent of any practical concerns, is critical to the advancement of science and the development of technology, the relevance of science is often measured in terms of its value to the end user. If the world is survival-oriented, then this last aspect will become increasingly more significant during next decades for geosciences in general, and physical oceanography in particular. Nowadays, the main societal concerns relating to physical oceanography are the humankind impact on the ocean and climate and the sustainability of the environment. Humanity is faced with pressing marine research problems, many of which arise from the need to accommodate multiple uses of the ocean and from the ever-increasing concentration of the population near the coasts. The demand for 'ocean products' from the society increases as it understands more about the anthropogenic impacts on the oceanic systems and the need for a better understanding of the ocean's role in controlling global chemical, hydrological, and climate processes. It can be expected that applied physical oceanography in support of socioeconomic planning and regulatory activities will become more significant as we move closer to the limits of sustainable use of the environment, and it should be emphasized that 'good science' and 'societally relevant science' are not necessarily mutually exclusive.

Physical oceanography has made giant steps in both directions. In the past decade, major breakthroughs were made in our understanding of ocean biology, chemistry, geology and physics. Such potential has been made possible by increasing capabilities in the areas of computation and rapid ocean-observing technologies. These new capabilities are providing unprecedented views of the ocean basins, their changing interaction with the atmosphere, the great biological shifts in near-surface waters, and the long-hidden deep ocean basins. Advances in ocean numerical modeling made possible by increased knowledge are going and will go on, greatly enhancing our understanding and predicting capability on time and space scales important to society. The promise of progress in the future is predicated on these new tools and the new "look" at the ocean that these tools have made possible.

4.3 Long-term goals for Physical Oceanography

Considering the concepts of the previous section, ocean scientists' ambitions and societal demands could merge in the following long-term goal for physical oceanography:

To attain a comprehensive understanding of the functioning of the ocean and its interactions with the atmosphere, land, cryosphere and biota in every temporal and spatial scale, and its response to natural and anthropogenic forcing to predict, manage, and modify its behavior to help sustaining the human race and our environment.

Even though our knowledge of the oceans and their role in the climate has substantially improved during the last decades, we are still far from fully reaching this goal that requires the construction of a valid 'Earth Model,' encompassing other geosciences in addition to physical oceanography. Besides the inherent limits of predictability (further discussed in the IAMAS Chapter 6), the extreme complexity of the system imposes technological and knowledge limits to a fast development of the desired predictability. The construction of such a model will require a much better understanding of all of the processes taking place in the coupled ocean-land-atmosphere-cryosphere system and their interaction with biogeochemical processes together with a much larger computing capability. Regarding ocean's role in such a model, ocean's physical, chemical and biological processes are intimately interconnected through a number of networks, interaction and interdependencies and feedback processes that must be fully understood prior to any attempt to modeling them and obtaining truly meaningful predictions. Meeting this scientific challenge requires research ships, incorporating new platforms of observation and sampling, acquiring long-term records of the relevant variables, developing new instruments, building a real time global observation system, improving theoretical knowledge, models and the ability to assimilate data and update forecast models, and increasing computer power.

Besides addressing the above goals, given the immediate societal demand related to environmental problems, physical oceanography should also identify intermediate goals. Two suitable aims for shorter time scales that meet great challenges to ocean sciences in general, taking advantage of the developing technology, providing important services to the society and constituting an intermediate step to the proposed long-term goal, could be:

- To develop a comprehensive predictive capability for the climate variations in a lifetime. The predictive capability should extend beyond foreseeing weather conditions and ocean temperatures to include effects of natural and anthropogenic climate change on sea life, including in fisheries and on land, which could, in turn, affect climate. The result of this work will eventually be a system capable of predicting climate fluctuations and their effects on the biosphere, both on land and at sea spanning in a lifetime.

- To develop a comprehensive understanding of coastal processes and their effects. As a result of this, we would expect to be able to deal with coastal problems without exclusive dependence on years of site-specific, routine measurements.

Our present knowledge and capability, the continued development in technology and progress in physical oceanographic research, promise that important advances can be done in this direction during the next decade.

4.4 Short term focus for Physical Oceanography

Reaching the aims proposed in the previous section requires development in almost every aspect and sub-discipline of physical oceanography, together with progress in meteorology, geochemistry and other geosciences besides a closer interaction among them. To be in position to reach those aims on shorter time scales, oceanography must improve our understanding of the ocean processes at every spatial and temporal scale. Some of the critical issues that should be addressed during the next decade regarding the two aims proposed in the previous section are outlined below. It must be emphasized that even though the global climate and coastal problems are analyzed separately, they are intimately interconnected, as they constitute two pieces of a unique 'ocean puzzle.' Therefore, progress in one of the areas depends on, and reciprocally will have important repercussion in, the other one. Moreover, many of the needs are common.

4.4.1 Climate and climate change prediction

Understanding how global climate evolves on decadal to centennial time scales is one of the most pressing environmental challenges that geosciences face today. Physical oceanography plays a huge role in unraveling the evolution of climate because of ocean's enormous importance in establishing the climate. Oceans contain 50 times more carbon than the atmosphere, dominate the hydrological cycle and, given water's large heat capacity, stores extremely large amounts of heat. These features give the ocean the capability of acting as a buffer, moderating both natural and anthropogenic forced changes in the climate system. Ocean circulation, through large-scale currents, thermohaline circulation, mesoscale motions and diffusive processes, plays an important role in distributing the heat, carbon and freshwater that is exchanged with the atmosphere, determining the climate to a large extent. The oceans absorb 40% of the carbon dioxide that is emitted into the atmosphere and are responsible for 50% of the polewards heat transport. As a result, the ocean sequesters heat and carbon in the deep waters and sediments.

Global climate prediction is probably the most difficult problem that our field has encountered. The dynamics are complex and involve a number of processes and scales. Their long time scales combined with the relatively short period of well-instrumented records of observational data make the direct observation of climate variability difficult or even impossible. Internal oscillations create climate variability significant enough to complicate and/or mask climatic variability produced by anthropogenic action. To assess the degree to which these forcing mechanisms are controlling our modern climate and the degree to which man has affected the climate system, the variability of the natural global climate must be understood. Given that we cannot wait to gather enough data to directly observe the climate variability, it is necessary to expand our data base and frame hypotheses about past climate change and ocean circulation using paleo-oceanographic studies and proxies, like ice cores, varved sediments, tree rings and corals (See Figure 4.1). Models are the essential tools for casting the results of theory and observations and for providing new hypothesis about the involved processes.

During the last decades, as a result of improved and extended observations, the development of pilot experiments and the evolution of numerical modeling and data assimilation techniques, there have been important advances in our understanding of the present and past climate systems.

One of the most important achievements has been the large improvement in observation due to a combination of new observational and measurement techniques and the development of global experiments. In-situ measurements of high accuracy made through experiments as WOCE (World Ocean Circulation Experiment http://www.woce2002.tamu.edu); CLIVAR (Climate Variability and Predictability http://www.clivar.org) and JGOFS (Join Global Ocean Flux Study http://www.uib.no/jgofs/jgofs.html) have given an unprecedented view of the contemporary ocean circulation, have created the first long time series of CO2 in the oceans and have allowed an improvement in the estimations of oceanic heat and CO2 transports. Polar orbiting satellites are providing a global coverage of critical properties every few days, like sea surface height (Topex-Poseidon http://topex-www.jpl.nasa.gov) and ocean color (SeaWIFS http://seawifs.gsfc.nasa.gov), that allow the study of the fluctuations is near-real time and at global scale.

The new measurement techniques have made possible direct observations of diapycnal mixing and basin-scale absolute velocity and wind fields and improved measurement and parameterization of air-sea fluxes, allowing for a better theoretical understanding of the ocean's role in climate and its modeling. The increasing sophistication of drifters and floats has allowed a view of the ocean circulation through Lagrangian pathways. The exploration under ice by sensors carried on autonomous vehicles permits us to survey the deep convection in progress, improving our knowledge and understanding of the water masses formation. Important advances have also been made in paleo-oceanography by improvements in the dating techniques and the development of new proxies. TOGA (Tropical Oceans and Global Atmosphere http://ioc.unesco.org) program has established the basic dynamics of ENSO and has lead to seasonal predictions with useful skill. Innovations in modeling and data assimi-

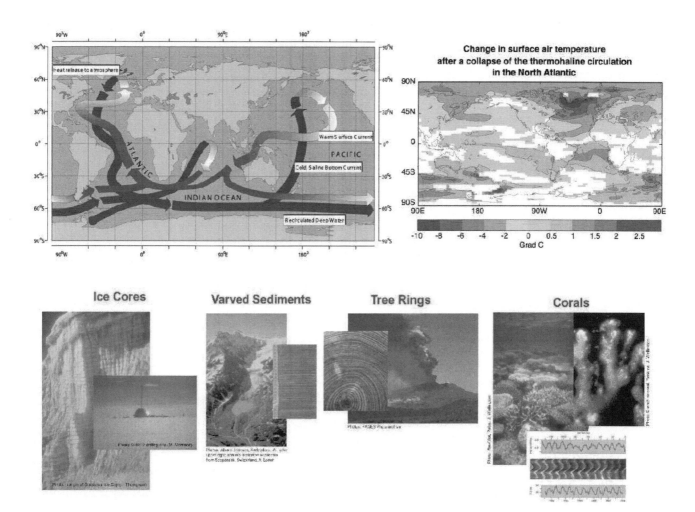

Figure 4.1: The ocean global thermohaline circulation (shown schematically in the upper left panel) maintains the present mild climatic conditions on the Northern Hemisphere. Upper right panel displays the change in surface air temperature after a collapse in the thermohaline circulation according to the Hadley Centre HadCM3 Model, showing an important cooling in that Hemisphere. Understanding this and other climate variability and climate change issues will require the application of realistic coupled models integrating all of the processes involved, the improvement of the observational base and a better understanding of ocean processes. Proxy data, as those shown in the pictures of the figure, are necessary to expand our data base and frame hypotheses about past climate change and ocean circulation Figs. courtesy of Clivar Project (http://www.clivar.org) and Vellinga, Michael and Richard Wood, 2002, Climatic Change, 54, 3 (251-267).

lation are including the full range of ocean biogeochemical cycles in models, leading to better predictions.

Innovative pilot experiments, like the Iron Fertilization (SOFeX http://www.mbari.org/expeditions/SOFeX2002) and the Brazil Basin Tracer Release Experiments (BBTRE http://www.whoi.edu/science/AOPE/cofdl/bbasin), pave the way for rapid advances to the understanding of CO_2 sequestration in the deep ocean that could reduce or eliminate the atmospheric residence time of this gas, mitigating the climate change.

Nevertheless, in order to reach the aim of developing a comprehensive predictive capability for the climate varia-

tions, a better understanding of the processes and couplings between the different parts of the climate system is needed. The field requires, therefore, further research and investment to develop, apply and evaluate more realistic coupled models that integrate all of the processes involved, the improvement of the observational base and a better understanding of ocean processes. Some of the main problems in this field, clearly interconnected, and several ways in which progress might be made are given in what follows.

Thermohaline circulation

The thermohaline circulation (See Figure 4.1) is one of the most important aspects of the ocean role in climate at long time scales. A quantitative understanding of these pathways and rates of the warm-to-cold conversion of water is crucial for both climate and climate change prediction. In order to understand the nature and extent of the ocean-induced delay in atmospheric warming created by rising CO_2 levels, determining the ocean's deep and intermediate water pathways for distributing heat and CO_2 and how these pathways influence the lag between atmospheric input and response is fundamental. An improvement of the understanding of the processes that control convection, affecting the thermohaline circulation's rate, volume and properties, is needed, particularly the role of fresh water. To reach this goal, a better understanding of the hydrologic cycle in climate is required. Knowledge of freshwater inputs from continents, precipitation, melting sea-ice, and an improved knowledge of the rates, pathways, storage and redistribution of water in all its phases are necessary. Understanding and quantifying the thermohaline circulation requires the study of a number of oceanic processes as meridional overturning, subduction, convection, western boundary currents and diapycnal mixing. The study of these processes is important as well as they control the geochemical recycling of nutrients.

Global freshwater circulation and sea-ice

As discussed above, the hydrological cycle is crucial for the thermohaline circulation, and it also has a direct impact on human activity and ecosystems. Unfortunately, data are too sparse to describe and properly understand the fresh water cycle, and models are still too crude to represent its different temporal and spatial scales of variability. A significant effort is required to observe and model the different aspects of the hydrological cycle and, in particular, of the fresh water fluxes between the different components of the climate system (land-ocean, land-atmosphere, ocean-atmosphere-sea ice). As an example, a key aspect that needs further development is the representation of sea-ice in numerical models. Because of its insulating capacity, its radiative properties and its creation of saline brine during freezing, sea-ice exerts profound effects on air-sea exchange. Its effect is obviously a major factor in high-latitude climate and has profound influence on the thermohaline circulation. The soon to become available satellite borne surface salinity measurements are expected to provide valuable information on the hydrological cycle.

Ocean mediated climate fluctuations

Our capability of predicting the dynamic evolution of the climate system depends upon our understanding of the decade-to-century climate variability of the coupled ocean-atmosphere system. Therefore, efforts must be made to understand ocean and ocean-atmosphere coupled modes of oscillation at those scales. Even though further studies are still required, the development of ENSO theories, observation, models and forecasting systems provide an example of a success story when international capabilities are joined in observing, understanding, modeling and predicting climate.

Other oceanic and climate phenomena would deserve similar international joined efforts. Examples are the Antarctic Circumpolar Wave, with a period of around 10 years and the North Atlantic Oscillation, with decadal periods. The strength and pattern structure and the coupled ocean-atmosphere dynamics of these climatic modes must be further understood. The role of greenhouse gases in forcing and modifying these modes should also be further explored.

The surface mixed layer

The surface mixed layer of the ocean is the one that directly interacts with the atmosphere, introducing into the ocean the wind and buoyancy fluxes. Misrepresentation of this layer can produce serious failures in predicting the onset, duration and depth of convective overturning events that set the properties of the subsurface water masses, and thus affect the thermohaline structure and the meridional circulation. Turbulence and flow structures of the scale of the mixed layer depth greatly complicate the physical behavior of flows (see Figure 4.2). Processes occurring at short time scales have, as well, an impact in this layer, as the diurnal cycle in heat flux that has a profound effect on the ocean surface, surface heating, and convection. Finding an optimal level of closure for mixed-layer problems represents a challenge to modeling. New instruments, like microstructure profilers, could optimally support the modeling activities.

Effects of complex seafloor topography and down-slope flows

Ocean circulation is very sensitive to bathymetry details. Seafloor mountains, ridges and continental rises exert horizontal pressure forces that are crucial elements of the circulation. Topography channels and breaks the flow into eddies that cannot be represented by moderate resolution models. As part of the meridional overturning circulation, water sinks in narrow streams or 'plumes' at high latitudes. As they descend, waters from intermediate depth layers are entrained into them, their transport increases and recirculating flows in the surrounding ocean are driven. Accurate representation of these concentrated flows in models is an important goal, which can be achieved as computer power increases.

Ocean turbulence

Ocean turbulence plays a critical role in the transfer of mass, momentum, salt, heat and other water properties from the larger scales of generation or input to the smallest scales of dissipation by molecular processes. Turbulence also has an impact on processes as diverse as zooplancton detection of prey and avoidance of predators, transfer of CO_2 across the air-sea interface, and mixing of heat and nutrients throughout the ocean. A better understanding of ocean turbulence and its role in ocean mixing, and how ocean mixing can be accurately parameterized is essential for developing new ocean circulation models for studies ranging from local ecosystem dynamics to coupled atmosphere-ocean dynamics.

Abrupt climate change

Paleo-oceanographic records show clear evidence that

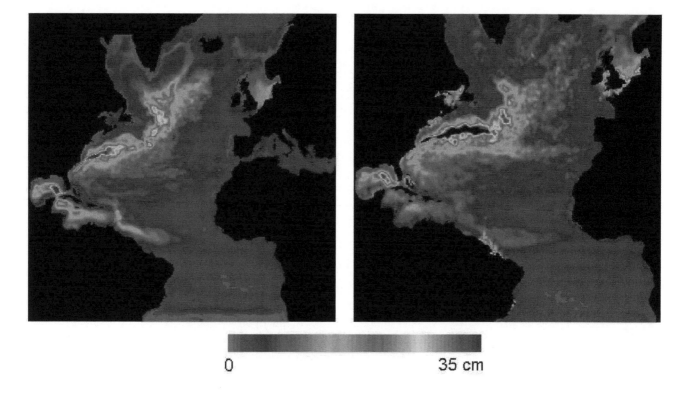

Figure 4.2: Sea surface elevation variability modeled by a high-resolution (1/12o grid), layered model of the Atlantic circulation with ECMWF daily forcing for the years 1979-1984 (left panel) compared to TOPEX / ERS-1 data (right panel). The high variability that characterizes oceanic circulation, increasingly better reproduced by models, is a central challenge to ocean modeling. Fig. courtesy of R. Bleck, MICOM Project (http://panoramix.rsmas.miami.edu/micom) and P.Y. LeTraon (Space Oceanography Division, CLS, Toulouse, France).

abrupt (within a few decades) climate change took place in the past, persisting for centuries. Even though the causes for these changes are still uncertain, the thermohaline overturning and the tropical circulation have been both suggested as possible mechanisms. Sophisticated climate models predict that changes in the hydrological regime, resulting from global warming, could reduce the rate of deep-water formation at high latitudes, producing a rapid collapse of the thermohaline circulation. This collapse would have impressive consequences for the global climate, like an important reduction of the mean temperature over Europe (see Figure 4.1). There are also indications that rapid releases of greenhouse gases to the atmosphere may have induced abrupt climate changes in the geological past. Therefore, it is important to determine the likelihood that such an event could result from anthropogenic activities. For that, it is necessary to understand what types of ocean conditions caused abrupt climate changes in the past, and what the consequences were to the ocean's ecosystems and the atmosphere. Given that two mechanisms for rapid change are known, and both of them involve the ocean part of the climate system, it is necessary to study whether other mechanisms have produced rapid changes in the past and whether those events could be repeated in the future.

CO2 sequestration

Techniques for carbon sequestration are a possible option to stabilize the greenhouse gasses concentration besides conservation effort and development of alternate energy sources. Given that the oceans are the largest active reservoir of carbon, they could be the appropriate repository for the excess that humans produce. Pilot experiments that either enhance the biological carbon pump or directly inject anthropogenic CO_2 into the deep ocean have already been successfully carried out. These options of carbon sequestration should be further explored together with the development of other techniques for mitigation.

Summary

Our predictive capability of the climate variations is currently handicapped by insufficient knowledge of the processes involved and the interactions between the different components of the climate system, by difficulties in representing some of those processes in numerical models and by insufficient observations in the required time and space scales. Nevertheless, with the observational database rapidly growing and with sufficient computational resources to support experimentation with a variety of models, we are on track to make rapid progress. One of the most important needs is the enhancement of our database. The importance and influence of satellite observations on research progress is evident, and therefore, it must be enhanced and further developed. Nevertheless, given that the ocean is virtually opaque to electromagnetic radiation, remote sensing from space is restricted to the ocean surface and its upper parts. Thus, the full understanding of the system requires sustained in situ observation. The very few existing long time-series stations capture the time domain at only one single point. New strategies for observing the appropriate spatial correlation are required. Deployment of a large number of new stations and efficient tools, such as instrumented floats and vehicles will greatly enhance our ability to observe and predict important planetary changes. Given that paleo-oceanographic studies are a key for understanding the evolution of climate, the efforts to extend the detailed records of oscillations in the ocean and climate system into the future should go hand-in-hand with efforts to extend them into the past. An increased knowledge of the oceanic fluxes of heat, water and biogeochemical properties that drive climate, together with a better understanding of the interplay of the internal oscillations of the deep and shallow ocean, are essential for evaluating climate change and for mitigation planning.

Large-scale numerical models of the ocean are the essential tools for casting the results of theory and observations into rigorous form. Ocean modeling, data assimilation and theory, therefore, must be continued, improved and employed. Fertile interaction of experiment, theory, data and modeling is, as well, necessary, in order to make significant progress in our understanding of the climate system.

Development within complex numerical models is particularly needed in areas of turbulent mixing and convection, complex sea-floor topography and down-slope flows, global freshwater circulation and ice and upper ocean dynamics. Models of biological and chemical processes are needed for prediction and understanding as, coupled with physical models, they allow exploration of ranges of response that can be realized, the conditions under which particular responses are possible, and thresholds where undesirable responses might occur. Finally, it is necessary to continue developing 'whole system' (Earth) climate models capable of predicting long-term climate change.

Deliberate tracer releases must be intensified in order to improve our understanding of the turbulent processes in the ocean that is crucial for a proper modeling of the climate system. The execution of perturbation experiments has resulted in important new information on the rate of ocean mixing, on fundamental controls on biogeochemical cycles and on the possibility of carbon sequestration. These techniques should be adopted, and new classes of perturbation experiments should be done, as they have the potential to give us the key to understanding climate and climate change processes.

4.4.2 Coastal oceans prediction

Even though the ocean margins occupy a relatively small portion of the ocean surface, its environmental importance is enormous. The coastal ocean (estuaries, continental shelves, and the Great Lakes) represents the part of the ocean that people most directly use, benefit from, and also affect. The coastal ocean is important for a number of economic reasons, including recreation, coastal development, mineral extraction, and fisheries. Indeed, most of the problems of sustainable ecosystems have a significant coastal component. This environment is being strongly modified and impacted by human activities, and an urgent need for better understanding the processes that influence and control it is emerging in order to provide cost-effective, accurate management advice.

In a broad sense, the great issue in the coastal oceans is how materials are transported and exchanged between the land, estuaries, the continental shelf, and the deep ocean, all the while undergoing physical, biological, and chemical transformations. This makes coastal oceanography extremely complex. The complexity of the coastal systems, their small scale and high spatial and temporal variability have been the main limits for understanding the processes that influence and control this environment. Nevertheless, the advent of new increasingly sophisticated sensing capabilities, the development of new modeling and data assimilation techniques and the continuous increase in computing power are creating the conditions to make an important advance in our understanding of the coastal systems, their prediction and management. The field requires further research and investment to improve the observational base, to understand the processes involved and their relations, interactions and feedbacks, and to develop, apply and evaluate more realistic models integrating all of the processes involved. A brief assessment of the main problems in this field and several ways in which progress might be made are given in what follows.

Physics of coastal environments

Physical processes strongly influence the character of the coastal ocean environment in a number of scales. Even though recent work gives a first-order understanding of many of the basic physical processes that act in coastal oceans, further research to gain a better knowledge of those processes will be necessary in order to overcome the barrier of predictability. Some of those processes are:

The role of turbulent mixing in estuarine physics:

Vertical mixing controls the estuarine circulation and the salt budget in most estuaries and plays a fundamental role in most coastal physical and biological processes. A more comprehensive and quantitative understanding of this process would improve our modeling and predictive capabilities. Direct measurements are needed in order to reach this aim. The new development of microstructure profilers is very promising in this context.

The behavior and dynamics of water plumes:

River plumes are one of the primary sources of freshwater, nutrients, pollutants and sediments of the coastal ocean. The study of the factors that influence the plume's movement and evolution over space and time and its interaction with the shelf and offshore or deep ocean currents systems is therefore extremely important. Understanding of these processes can be only reached through further observations and application of numerical models.

The dynamics of the surf zone and inner shelf:

Understanding the dynamics of the inner shelf is not only locally important because of, for instance, its influence in the coastal morphology, but also because of its important implication and the along and cross-shelf transports. Further field studies are required to understand those processes and validate numerical model results.

The dynamics of wind-driven currents:

The knowledge of the shelf flow regime and the off-shelf transports is central for coastal applications. These processes are strongly influenced by details of the bathymetry and wind forcing on very small scales. Further development of high-resolution numerical models and high-resolution observation of the forcing and topography are needed to address this problem.

The bottom boundary layer:

Due to the shallowness of the estuaries and shelf, turbulent drag and mixing generated by near-bottom flow play a central role in controlling many processes. Biota also has an effect on the bottom drag given that benthic biology modifies the bottom roughness. A better understanding of these processes and their parameterizations is needed to improve numerical model's predictions. Laboratory based studies and increases in observational capabilities would improve this subject.

The shelf break fronts:

Even though the shelf break fronts are features present in many coastal oceans, not much is known yet about them. The detailed thermohaline and velocity structures of the shelf break fronts and shelf break jets need to be quantified, and the impact of these processes on material transport, the spread of marine organisms and the formation of geobiological domains needs to be understood.

Physical forcing and biological processes

While some effects of physical forcing have been established, there is a great uncertainty in the mechanisms by which they influence variability in biological phenomena, even though it is known that this effect can be dramatic. Biology can have, in turn, an effect on physics, for example, due to the aforementioned modification of the bottom roughness. Assessing the effects of physical processes and their natural and anthropogenic variability on coastal biological communities is a problem of clear ecological and economical importance that demands further efforts.

Land-Ocean fluxes

Material inputs to the coastal zone from land include organic and inorganic sediment, dissolved elements, dust and anthropogenic pollutants and nutrients. These inputs are influenced by the local watershed, meteorological characteristics and biogenic habitat quantity and quality. These material inputs have a major influence on the chemistry of the global ocean and are greatly influencing ecosystem processes, including food web dynamics, biological production and community structure. Some important issues that require further study include:

Inputs of biologically important materials:

Even though the water flow and major element composition of larger rivers are reasonably well known, the chemical form of minor elements and organic matter that are transported from terrestrial watersheds to coastal environments is less understood. Riverine oceanic matter is fundamental in the global carbon balance, and the minor elements are often critical to coastal and oceanic ecosystems. The groundwater discharge can locally influence, in a significant way, the freshwater contributions and often, has higher concentrations of dissolved species. Efforts must be made to identify regions of groundwater influence, its role in material flux and consequent ecosystem response.

Sediment fluxes:

Globally, rivers deliver around 20 billion tons of suspended solids to the sea every year. These sediments are not chemically inert and may serve as a source or sink for nutrients, pollutants and other materials in the coastal ocean. While some aspects of sediment resuspension and transport are understood and supported by field measurements, the conceptual and numerical modeling and prediction of sediment transport requires further development. Bioturbation is a key process, as benthic organisms distribute the deposited material over a certain depth of the sea bottom, which in turn influences the erosion processes. There is a need, therefore, to investigate the coupled system suspended particulate matter dynamics-biology-morphodynamics. This is a fundamental need in order to construct confident budgets and to predict the evolution of the shoreface and bottom topography under different conditions.

Carbon fluxes:

Margin sediments can be efficient in sequestering carbon, and this preservation could influence the CO2/O2 balance of the atmosphere on geological time scales. Nevertheless, the mechanisms that control this sequestration are not well understood. Organic carbon buried in sediments may serve as a substrate for microbial activity within the sediments, giving rise to gas hydrates in ocean margins. Given the enormous size of the reservoirs, these hydrates may have significant influence on the carbon cycle and the climate. Their distribution, nevertheless, as well as their generation rates, migration pathways through the sediments and rate of return to the ocean environment are poorly known.

Anthropogenic influences:

Deforestation, agriculture, flow modification, coastal constructions, urbanization and other human activities dramatically influence rivers, watersheds and coastal sediment transports by changing the amount of water that is transported and its chemistry. For instance, global river discharge of nitrogen and phosphorus and total organic carbon have doubled during the past two centuries. Together with fishery exploitation, it has produced the largest impact on biological processes in coastal ecosystems. Human activity injects into the atmosphere particles, aerosols and chemicals. The amount of the dust flux, its variability and its influence on biological systems through delivery of contaminants and limiting micronutrients is not clear. Similarly, biological communities are affected by other chemicals of anthropogenic origin released to the ocean, like trace elements and toxics. Greater research is needed to determine the ecological effects of human and natural alterations in the quantity, quality and spatially explicit processes associated with coastal habitats. Efforts are also needed to identify and predict how global warming and sea level rise will affect the distribution, abundance and landscape connectivity of various habitat types and their influence on associated processes and species. Interdisciplinary research between physical oceanography, biology and chemistry, involving modeling and observations integration will be necessary to address these questions.

Exchanges between the Deep Ocean and Shelf Ecosystems

The connections and exchanges between the small-scale processes of the shelf and the large-scale of the ocean are increasingly becoming an important oceanographic issue, as they can have a global impact. Localized export of material, for instance, can have global importance as it is transported far by ocean currents, changing the chemical properties of distant environments. Other examples are the growth of plankton populations, which affect carbon dioxide levels, and thus may be important in global warming scenarios, the anthropogenic pollutants that have reached the open ocean, which are known to be transported far from their sources and the nitrogen global cycle, which is strongly dependent on denitrification that takes place on the ocean margins. A better understanding is needed of small-scale processes and small-scale aqueous systems (estuaries, wetlands, coral reefs) to understand these exchanges and evaluate their impacts on global issues. Quantifying the exchanges will be required to determine the magnitude and time-dependent nature of shelf-ocean and sediment-water fluxes in interdisciplinary efforts with geologists, biologists, hydrologists and biogeochemists.

Summary

The ocean margin occupies only about 10% of the global ocean area, and a much smaller percentage of its volume, but it dominates the economic importance of the sea. Material fluxes to the coastal zone are largely unidirectional, from the land to the sea, and these fluxes of sediments, biologically active materials and sometimes toxic compounds are rapidly increasing. If we are to comprehend the impact of these fluxes, we must conquer the complexity of coastal

transport processes. The physical forcing of wind provides coastal upwelling, which enriches marine ecosystems and creates rich fisheries. Natural climate variability can have strong impact in these systems, as it happens with El Nio, for example. Other natural or anthropogenic climate shifts could change these dynamics with enormous consequences. In order to solve these and other important coastal problems, advanced technologies in sensors and observing systems, interdisciplinary research to understand the processes and their connections, integrated numerical models and data assimilation are needed.

New eyes from satellites and coastal radar can provide needed observations in the small time and space scales inherent to many coastal processes. Even though the time and space scales of the coastal processes can be small, they present large variability in interannual scales. Therefore, there is a need of monitoring them over long periods of time. Observing the system over an extended time will allow for a better determination of how the system responds to major natural changes, enhancing our capability of predicting its behavior and its response to anthropogenic perturbations.

In studying the diverse roles of physical processes in other coastal phenomena and the local ecosystem dynamics, accurate in-situ observations play a fundamental role. The continued development and use of moored instrumentation in physical oceanography, biology and chemistry will result in major advances in a number of important areas as biophysical coupling, cross-shelf transport processes, turbulent mixing processes and their parameterization in numerical ocean models, buoyancy-driven flows and fronts, sediment transport and biogeochemical fluxes.

Turbulent mixing processes are critical to many coastal processes. So, continued emphasis should be placed on field studies of the turbulent cascade of energy. Improvements in our knowledge of turbulence, for example by means of newly developed microstructure profilers, will result in an important improvement of our modeling capabilities. Besides their intrinsic interest, these studies provide the basis for comparison with new numerical studies of these proces8ses and the development of better parameterizations of turbulent mixing. This last, in turn, would benefit both coastal ocean research and the use of models for practical applications.

As mentioned above, modeling plays a key role in coastal oceanography. Operational real-time regional ocean prediction systems (see Figure 4.3) are required to support critical activities in the coastal environment, like navigation and marine operations, oil and pollutants dispersion control, search and rescue, prediction of harmful algal blooms and other ecosystem or water quality phenomena. These systems require, in turn, to be supported by operational oceanographic monitoring of oceans.

Realistic interdisciplinary models in concert with data assimilation are an essential tool for improving our understanding of complex coupled biogeochemical and physical

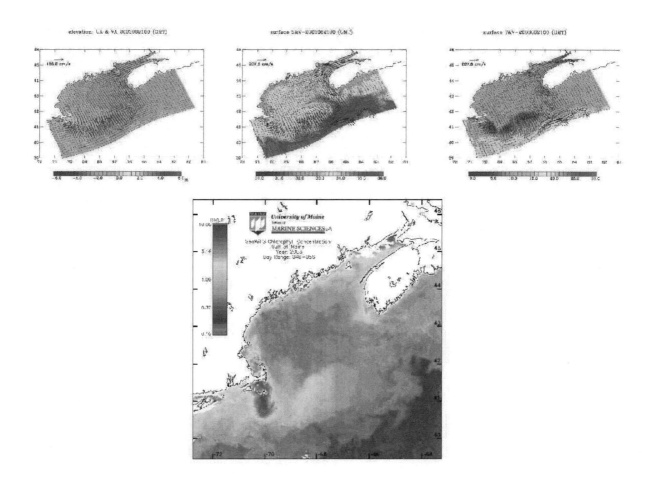

Figure 4.3: Coastal models can be valuable as management tools using monitoring and data assimilation. Figure displays examples of ocean model predictions (upper panel) of surface elevation and currents (left), surface salinity (center) and surface temperature (right), and satellite color observations (lower panel) from GoMOOS (Gulf of Main Ocean Observing System http://www.gomoos.org). This kind of systems, that bring together managers, industry, researchers, educators and others in a collaborative effort to establish a large-scale regional ocean observing system, can enhance, facilitate, and protect the livelihoods and well-being of those who use the water bodies and depend on their ecosystems.

processes in coastal areas. Besides, given that biological and chemical data sets are sparse, it will permit to extend the database through a dynamical interpolation providing a framework for detailed diagnosis of those processes. The instrument and platform development occurred during the last years together with the rapid evolution in numerical modeling, the development of theory and the continuous progress in communications are promising an important evolution of this field during the next years.

The need for a synoptic view of this environment to understand the processes that produce changes in coastal systems presents major challenges for coastal oceanography. New couplings of chemistry, biology and physics will be necessary to develop a better sense of them.

4.5 Recommendations

Taking geosciences to a point where answers will be given to many of the urgent environmental questions of this new century will demand effort and commitment of both scientific community and policy makers. Physical oceanography plays a crucial role in the development of the 'Earth system science', required to meet this challenge, as the oceans are a central piece of the puzzle. The field itself requires a better understanding of the ocean's behavior, its connections, interactions, feedbacks and couplings. Therefore, physical oceanography needs further research and investment to develop, apply and evaluate more realistic models integrating all of the processes involved, the improvement of the observational base, a better understanding of ocean processes and a much larger interaction between ocean scientists. But there will be, moreover, a need for further integration among geosciences to meet the real challenge of understanding and predicting the environment for properly managing and sustaining it in a multi-coupled system. On the other hand, the anthropogenic influence can no longer be neglected in an integrated model in the long-term. The impact of, for example, fisheries, coastal constructions, like dikes, harbors wind farms, etc., and river management, has such a strong effect on the coastal system that the governing socio-economic drivers must be taken into account; if it is intended to predict such a system.

In what follows, a summary of the physical oceanography science needs and strategies for achievement emerging from this report, which can be regarded as a few recommendations to the oceanographic community and policy makers, are highlighted.

Global Observing System

Ocean observations have always been the driver of new knowledge and predictive capabilities in the ocean and its basins. Despite this, the oceans remain vastly undersampled in time and space. Important progress in this aspect has been derived, and will continue on, from the development of new observational satellite tools. Nevertheless, given that remote sensing from space is restricted to the ocean surface and its upper parts, sustained observation in the ocean is required. An international effort to support sustained high-quality global observations over decades must be done. Measurements of oceanographic, meteorological, hydrologic and biogeochemical variables are necessary to meet new scientific challenges and practical needs.

The need for a strong Global Ocean Observing System (GOOS http://ioc.unesco.org/goos/) as a component of the Global Climate Observing System (GCOS http://www.wmo.ch/web/gcos/gcoshome.html) is clear. The information to deal with marine related issues is needed, by scientists, governments, industry, and the general public. This includes the effects of the ocean upon climate. A unified global network to systematically acquire, integrate and distribute oceanic observations is required and must be supported.

New observing techniques and instrumentation

The success of new observing techniques and instrumentation in revealing aspects of the ocean circulation and climate has been spectacular over the past 30 years. Good examples are the recently developed microstructure profilers that allow the observation of, for instance, the dissipation rate, and the ARGOS system which collects environmental data from autonomous platforms and delivers it to users worldwide. Improved observing techniques must continue to be developed both to make more effective measurements of essential variables and to expand the suite of variables that can be observed.

Data assimilation

Data assimilation can add considerable value to global observing systems by combining diverse sets of observations with global numerical models to produce comprehensive and internally-consistent fields. The assimilated products combine the accuracy of the observations with known geological, chemical and physical consistency and provide global coverage inherent in the models. Therefore, advancement of assimilation techniques to reach optimal schemes must be a priority.

Operational oceanography

An operational oceanographic near-real-time data collection, similar to the one existing for meteorology, needs to be established. There is a need for efficient, integrated systems for coastal and global ocean products, provided regularly and in real-time. That system, which requires the application of state-of-the art models, data assimilation methods, global observing and communication, would have tremendous impact on improving coastal, open ocean and climate forecast, as well as on most of the aspects of physical oceanography research. Experimental systems like GODAE (Global Ocean Data Assimilation Experiment http://www.bom.gov.au/bmrc/ocean/GODAE, see Figure 4.4) must, therefore, make the transition from research mode to operational GOOS and CGOOS (Coastal Global Ocean Observing System http://ioc.unesco.org/goos/cozo.htm).

Numerical modeling

A continued emphasis is needed on the development of models in all their applications, and a special effort is needed in developing models that link the different parts of the ocean system.

Numerical models are becoming a centerpiece of our science since results of theory and observational data are often cast into the form of numerical models. This happens either through data-assimilation or through process-model explorations of theoretical ideas. Models have proved essential in determining the relative importance of variables in complex, nonlinear systems that have multiple feedbacks and are the ultimate tool for prediction. Crucial physical effects are being now identified that have seriously limited the accuracy of model prediction in the recent past. By taking advantage of rapid advances in computer technology, and of new data streams from powerful observing programs, great advances will be made. There is a need, therefore, for model and data assimilation, development and improvement. Linked models that tie together physical, geological, biological and/or chemical systems show great promise and should grow in importance in oceanographic research. Interactions between such modeling effort and field programs will aid in both designing observation strategies and in evaluating outcomes.

Closer interaction among physical oceanographers

Overcoming the predictability barrier will require the continued development of in-situ and satellite observing systems, improved numerical models, the assimilation of data in models, and the closer linking of the physical system to the important biogeochemical aspects of the ocean and climate. Reach-

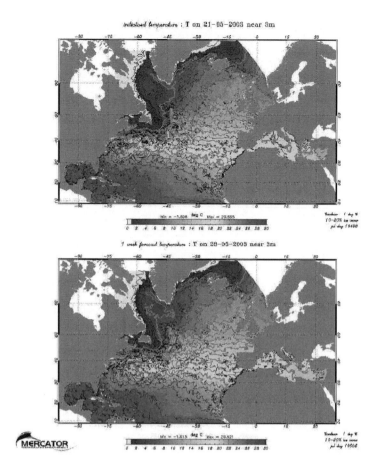

Figure 4.4: Sea surface temperature nowcast for 5-21-2003 and one week forecast from MERCATOR mission. From 2003 to 2005, the demonstration phase of the GODAE experiment will provide a routine and real time depiction of the global ocean. Half a dozen of operational oceanography centers will emerge in this purpose throughout the world. MERCATOR will be one of them. The transition from the experimental phase to a truly operational one would have a tremendous impact in physical oceanography. Fig. courtesy of MERCATOR Mission (http://www.mercator.com.fr).

ing these objectives will, therefore, demand a close interaction between seagoing oceanographers, scientists concerned with the ocean-atmosphere interface, ocean modelers, theoretical oceanographers and satellite remote-sensing scientists that could be promoted by improving the communication and interaction among scientists of different fields and countries. IAPSO might play a fundamental role in this aspect. Some suggestions are:

- A greater and better use of the Internet.

- Workshops that bring oceanographers from different institutions and areas of expertise together for extended periods.

- Opportunity for investigators to make extended visits to institutions other than their own and to meet others with similar interests, but different approaches.

- Summer schools, workshops and post-doctoral programs to encourage cross-fertilization between the different methodologies.

- Encouragement of proposals using complementary approaches (observations, modeling, laboratory, theory).

Earth System Science and interdisciplinary research

As oceanography progresses and faces more often environmental problems, it becomes increasingly necessary to formulate new research initiatives using an interdisciplinary framework. We should facilitate closer ties and collaboration across the geosciences.

To meet the aim of building up an 'Earth Model' for understanding and predicting the functioning of the whole climatic system, collaboration of experts in the traditional earth disciplines (meteorology, oceanography, hydrology) as well as in ecology, information technology, instrumentation and complex system theory will be necessary. In this new science, the

40

most challenging task is integration, as we need a vision as integrated as possible of all the many elements of the system. Physical oceanography must make commitments to reach the degree of co-ordination and co-operation among geosciences demanded by this enterprise. The IUGG and its comprised associations have a fundamental role to play here.

Public information and communication:

The ocean plays a critical role in our everyday life and in the future of our planet. As more people understand this and begin to appreciate the Earth as a water planet, they will take action to conserve the ocean and the web of life it supports and will give, consequently, a better support to oceanographic scientific research. Therefore, the oceanographic community should make an effort to improve presentation and communication of results to society and policy makers, taking special care of avoiding misinformation and/or poor use of the information. This should not be a one-way road. In the frame of integrated models, oceanographers must also account for social and political decisions if long-term predictions are envisaged. Communication is an important aspect in which IAPSO can play a central role.

E-science:

A greater and better use of the possibilities of free information and communication provided by the Internet would be of great benefit for geosciences.

Infrastructure:

In order to reach our aims, we need adequate updated technology. The availability of research ships requires inversion and planning. Continued enhancement of computing resources is required in order to meet the demands for climate modeling and forecasting. Even though the power of computers is doubling every few years, the complexity of the oceans ensures that the needs of physical oceanographers will always be limited by computational resources. Physical oceanographers need continued access to computing resources, which are comparable to those of the international atmospheric and engineering communities.

Human resources:

If we are to reach the goals, there must be sufficient scientists with an adequate education researching all over the world. On the other hand, we must assure that a steady supply of experimental scientists and engineers continues along with modelers and theoreticians in order to keep the balance between the different fields. As environmental problems will increase in the next decades, future graduates should have increased opportunities for interdisciplinary studies both inside of and outside of the academic science community. Human resources can constitute a serious problem in developing countries, where the low number of experts (and sometimes their lack) impedes the formation of new graduates. The degree of global expertise could be greatly enhanced by the creation of inter-institutional PhD programs promoted by IAPSO.

4.6 Conclusion

Physical oceanography has reached a challenging point where progress in technology and the tools that this technology make possible are allowing for a new 'look' at the ocean. The complexity of the ocean circulation may now be approached with sufficient detail to promise that significant progress will be made on many scientifically and societally important issues during the next decade. IAPSO and the oceanographic community should take advantage of this situation.

The great challenge of constructing an 'Earth Model' to improve our understanding, use and management of the Planet for the benefit of the Human Race, in which physical oceanography will play a fundamental role, can start to truly be faced. It will become a reality if the necessary resources are available and a genuine commitment is made by the geosciences community to reach the required level of integration and research effort. IUGG, through its actions, can play a significant role in the realization of this vision.

4.7 Acknowledgments

The GTF committee would like to thank Alan A. Meyer for his work on the IAPSO section of the report, and serving on this committee for the amount of time that he was able.

Bibliography

50 Years of Ocean Discovery: National Science Foundation 1950-2000. Commission on Geosciences, Environment and Resources (CGER), Ocean Studies Board (OSB), National Academy Press. Washington, D.C. 2000.

Coastal Ocean Processes and Observatories: Advancing coastal research. Coastal Ocean Processes (CoOP) Report No. 8. Technical Report. 2002.

Current Approaches to Seasonal to Interannual Climate Predictions - IRI Technical Reports 00-01, 2000

IPCC, Climate Change 2001: The scientific basis. Technical report, IPCC, 2001.

Global Environmental Change: Research Pathways for the Next Decade, Committee on Global Change Research, National Research Council. National Academy Press. Washington, D.C.. 1999.

Global ocean science: toward an integrated approach. Ocean Studies Board, commission on Geosciences, Environment, and Resources, National Research Council. National Academy Press. Washington, D.C., 1999.

GODAE (Global Ocean Data Assimilation Experiment), 2001: GODAE Strategic Plan, GODAE International Project Office.

Ocean Circulation and Climate. Observing and Modelling the Global Ocean. G. Siedler, J. Church and J. Gould Eds. Academic Press. London. 2001.

Ocean Sciences at the New Millennium Report. Workshop sponsored by the National Science Foundation through an award to the University Corporation for Atmospheric Research Joint Office for Science Support. Edited by Geosciences Professional Services, Inc. 2001.

Oceanography in the Next Decade. National Research Council, Commission on Geosciences, Environment, and Resources, Ocean Studies Board. National Academy Press, 202 pp., 1992.

Opportunities in Ocean Sciences: Challenges on the Horizon, Ocean Studies Board, Commission on Geosciences, Environment, and Resources, National Research Council, 1998.

Report of the Advances and Primary Research Opportunities in Physical Oceanography (APROPOS) Workshop, Monterey, California, December 15-17, 1997.

Strategic Plan and Principles for the Global Ocean Observing System (GOOS). GOOS Report N° 41. IOC/INF 1091. 1998.

Exploration of the Seas: Interim Report. Ocean Studies Board (OSB). National Academy Press. Washington, D.C. 2003

The second report on the adequacy of the Global Observing Systems for Climate in support of the UNFCCC. GCOS-82. WMO/TD No. 1143, 2003.

The integrated, strategic design plan for the coastal observations module of the Global Observing System. GOOS Report No. 125; IOC Information Documents Series N1183; UNESCO 2003

Theory in Ocean Dynamics. P. B. Rhines. In The Future of Physical Oceanography, report of the APROPOS Workshop, Monterey CA UCAR/NSF, 178pp. 1998

Towards Life and the Whole Earth; Ocean Science New Directions. Japan Marine Science and Technology Center, 2002. 19

Chapter 5

International Association of Hydrological Sciences

T. Oki

5.1 Introduction —Hydrology, the forgotten earth science—

There was a statement from Prof. R. Bras at MIT [*Bras and Eagleson*, 1987] in the 2nd US-Japan Hydrology Seminar in Hawaii [*Kayane*, 1989]:

> Hydrology, the science of water, has natural place alongside oceanography, meteorology, geology, and others as one of the geo-sciences; yet in the modern science establishment, this niche is vacant. Why is this?

Such an awareness of the issues has been discussed repeatedly in various workshops and symposia and even published in a few books [*Kundzewicz et al.*, 1987; *NRC*, 1991; *Buras*, 1997].

One of the answers to the above question is that hydrology is too biased to engineering:

> In practice, hydrology is regarded mostly as a technological discipline rather than a science. This attitude is responsible for much poor science in hydrology, which, in turn, has led to much poor technology in applied disciplines. It is urged that the present offers a unique opportunity to fill the vacant niche of the science of hydrology and some suggestions for action are offered. [*Klemeŝ*, 1988]

Until recently, engineers in hydrology were forced to give some account for demanding and sometimes unreasonable questions from society, based on limited and uncertain information of water balances and hydrological cycles. Scientists in hydrology tried to pursue "pure science" but not a few of them failed to overcome the temptation to relate their research with practical applications.

Here, the clue is that "pure science" and "practical engineering" should not always be exclusive, at least not in hydrology. Before the era of global environmental issues, human activities were excluded from the view of natural sciences. However, anthropogenic effects should not be excluded when the current Earth system is discussed, since "natural" and "real" have different meanings in this Anthropocene [*Crutzen*, 2002]. The "real" situation of the Earth is not "natural" anymore.

Hydrology should have the leading potential among geosciences to explore the interactions between human society and natural earth systems. The current struggle of hydrology to realize potential achievements is reviewed, and the perspectives of hydrology are presented in this short report.

5.2 Long-term Goals

5.2.1 What is hydrology?

UNESCO defined the discipline in 1964 as:

> Hydrology is the science which deals with the waters of the earth, their occurrence, circulation and distribution on the planet, their physical and chemical properties and their interactions with the physical and biological environment, including their responses to human society.
>
> Hydrology is a field which covers the entire history of the cycle of water on the earth.

It is noteworthy that the responses to human society are included in this statement issued just before the International Hydrological Decade, which was the first international research program on hydrology, organized by UNESCO for 1965-1974.

The recent definition [*UNESCO and WMO*, 1992] says hydrology is

1. Science that deals with the waters above and below the land surfaces of the Earth, their occurrence, circulation and distribution, both in time and space, their biological, chemical and physical properties, their reaction with their environment, including their relation to living beings.

2. Science that deals with the processes governing the depletion and replenishment of the water resources of the land areas of the Earth, and treats the various phases of the hydrological cycle.

The two parallel definitions are from the natural science side and from application oriented side.

Then, which water and what hydrological cycle are excluded from the view of the current hydrological science? These are for example:

- water out of the Earth: even though water is not unique on this planet.

- water in the Earth: even though water may have a significant role in the motion of the mantle, and the total amount included in the Earth could be more than that on the surface of the Earth.

- ocean: even though more than 97% of water on the Earth's surface stays as oceanic water.

- atmospheric water: even though there are certain amounts of studies on precipitation by hydrologists and the discipline of hydrometeorology exists, generally precipitation is the given input to hydrological system on the land surface.

- water in pipes: water cycles in purely artificial system is not the main target of researchers in hydrology.

- organic water: even though water flow in plants is intensively examined in current hydrology, water cycle in circulatory organs is not generally dealt with by hydrologists.

From the above list, it is apparent that hydrological science covers the niche of other disciplines by nature. On the contrary, transdisciplinary studies in these currently uninvestigated water cycles by hydrologists may have some potential to break through the new frontier in hydrological sciences.

5.2.2 Long-term Goals in Hydrology

The ultimate objectives of the hydrology are to know:

- the quantity and quality of water in a defined area, and

- the inflow and outflow of water and its content to and from the area.

It could also be relevant for hydrology to utilize the knowledge of water for society and human beings.

Therefore to increase the ability to:

- monitor the quantity and quality,

- model/simulate the flux, and

- predict the quantity, quality, and flow of water as required to promote the hydrological sciences.

All these monitoring, modeling, simulations, and predictions should be associated with the estimates of their uncertainties and reduction of the uncertainties is one of the targets in hydrological science.

Similar to other geo-sciences, finding another principal law governing the movement of water is not the central theme of hydrological science. However, it does not mean there is no opportunity that new laws in hydrological science have been/will be found. Because of the scaling effect and the heterogeneity of the boundary condition and/or environment of the phenomena, theories and equations, which can be applied to ideal, well controlled conditions, either molecular scale or laboratory experimental scale, may not have enough capability to explain the phenomena we observe in the real field. This demands the establishment of sophisticated systems of laws and theories that can describe the hydrological cycles on the Earth, considering the spatial and temporal scales and the heterogeneity.

In other words, consideration of scaling effects and heterogeneity is identifying the dominant process governing the particular hydrological phenomena. In a sense, hydrological science formed its original discipline apart from fluid dynamics and hydraulics by neglecting the conservation of momentum by defining the target phenomena with viscosity prevails the motion and cancels the acceleration. Future hydrological science may include the detailed equations of Motion, if needed, for better simulation/prediction of hydrological Cycles. However, full descriptions of detailed hydrological processes in the real world, even for a hill slope of 1m length, will remain impractical and may be useless for understanding the phenomena.

Here, "full description" indicates describing the phenomena based only on the principal physical laws with reliable parameters, which can be estimated in the laboratory or field experiments. The value of the parameter should be reproduced with enough accuracy when the experiments are repeated, and it is preferable the parameter can be estimated from other physical parameters.

In this sense, even Darcy's law, which is one of the basic laws in hydrology, is not the principal physical law. Darcy's law requires saturated hydraulic conductivity as its parameter, but it is well known that saturated hydraulic conductivity is various even within a one square meter field, which seems homogeneous in terms of the density and the contents of the soil. However, it should be still better for most of the problems in hydrology to adopt Darcy's law rather than considering water flow through each microscopic pore in the soil layers. Therefore, integrated theories/models/equations corresponding to the temporal and spatial scales and heterogeneity will be used to describe the phenomena.

5.3 Short-term Goals

5.3.1 Current Demands on Hydrology

Conservative projections, based on current information about population growth rates, pollution, salinization, waterlogging and groundwater overdraft, all indicate a high increase in the number of countries that will experience significant water stress and scarcity in the near future. The United Nations system presented a Comprehensive Assessment of the Freshwater Resources of the World [*UN et al.*, 1997] to the nineteenth Special Session of the United Nations General Assembly in 1997. This assessment made it clear that a "business-as-usual" approach to water allocation and management is not sustainable.

The situation for a majority of the world's population has not improved since the Comprehensive Assessment of the Freshwater Resources of the World was presented to the United Nations in 1997. Still, over a billion people lack access to safe drinking water, and almost 3 billion lack proper sanitation. Millions of people die each year from water related diseases, and there are no clear indications that these trends will improve in the near future. The increased use of pollutants associated with the technological development of countries with evolving market economies, life-style changes all over the world and the increased need to develop sustainable water resources and infrastructure for growing populations all contribute to the urgency associated with advancing hydrological sciences and water management strategies. Increased competition among and within sectors calls for strategies that will help to balance water use for the multitude of purposes required by humans, while protecting the integrity of the ecosystems. Some of the statistics are staggering. By 2020, water use is expected to increase by 40%, and 17% more water will be required for food production to meet the needs of the world's growing population.

In the millennium declaration (55.2), which was adopted by the General Assembly of the United Nations in September 2000, it was stated in section 19;

> To halve, by the year 2015, the proportion of the world's people whose income is less than one dollar a day and the proportion of people who suffer from hunger and, by the same date, to halve the proportion of people who are unable to reach or to afford safe drinking water.

A similar statement was included in the Plan of Implementation from the World Summit on Sustainable Development held in Johannesburg in August 2002, in article 6 of Chapter II Poverty Eradication. In addition to that, the access to basic sanitation was also focused in article 7. In the WSSD Plan of Implementation, further research and development on water related sciences was urged together with researches on energy technologies, bio-technology, environmental management, marine resources, and sustainable development. All of these international vision/action statements aim to alleviate Poverty, particularly in developing countries.

Under such a movement, what kind of researches can be offered from the scientific community in order to contribute to solve the current water related issues and the anticipated "water crisis" in the near future in the world? What are the scientifically excellent and practically/socially relevant research topics?

Over the last decade it has become increasingly clear that if we are to face the water challenges of the future, we must view the Earth as a single, though highly complex, integrated system. The atmosphere, the hydrosphere and the biosphere extend throughout the globe. They are very dynamic and highly interactive. Changes that occur in one location can influence the environment somewhere else in ways that are not always expected and are frequently poorly understood. The complicated interactions between the physical and human systems are equally critical. The scientific and technical community has a very important role to play in this regard. This role is not only related to scientific monitoring, analysis, and interpretation, but also to guiding future policies and action programs. Effective water management is not purely a political problem, a scientific problem, a technological problem, nor a managerial problem; it is a complex problem that demands interaction between all of these areas. It is clear that to assuage the impending water crisis, cooperation between the scientific disciplines, as well as the scientific and water policy arenas, must dramatically improve. A directed effort must be taken now to improve the recognition of the impending water crisis by all stakeholders and to improve the understanding of the components involved with the natural system and allocation of its resources.

5.3.2 Contribution to solve the world water crisis

In preparation for WRAP (Water Resources Application Project) under GEWEX (Global Energy and Water Cycle Experiment) Hydrometeorological Experiments Panel (GHP), these items were identified to contribute to solve the current water related issues. These are to provide

- reliable information on

  - past and current global hydrological cycles,

- near-real time to seasonal prediction on natural variability of climate, such as El Niño and La Niña, and

- the change in hydrological cycles under climate change conditions, such as global warming,

• quantitative estimates of how much water is (and will be) really available for further water withdrawal for human beings, and

• alternative measures to be taken in order to secure the water supply to meet the water demand.

The first three items under the category of "provide reliable dataset" can be done by natural scientific approaches, however, the other two should require inter-disciplinary collaboration between natural and social sciences. The planned Global Water System Project by WCRP (World Climate Research Programme), IGBP (International Geosphere-Biosphere Programme), IHDP (International Human Dimensions Programme), and DIVERSITAS (An International Programme of Biodiversity Science) will be an appropriate platform to pursue such a collaborative research (http://www.gwsp.org/).

To provide a reliable dataset of past and current global hydrological cycles, continuation of observational network, research and development on new observational measure, and "data rescue" of historical records are important. Satellite remote sensing could be the only hope to provide global coverage of hydro-climatic information, however, ground based network should be held by all means to secure the reliability of global observation. Since there are not a few historical records that are not digitized but cover more than hundred years, database development of these data is also important.

Global offline simulation of LSMs (land surface models) can provide global flux and balances of water, energy, and even more (such as carbon). However, the accuracy of the estimates is subject to the accuracy of LSM itself, forcing data, model parameters, and temporal and spatial scales employed. Therefore the uncertainties associated with the global estimates by LSMs will be examined in the GSWP (Global Soil Wetness Project) under GEWEX/GLASS, Global Land Atmosphere System Study (http://hydro.iis.u-tokyo.ac.jp/GLASS/).

There are many researches to improve providing seasonal prediction of water cycles. It seems widely believed that the improvement in giving boundary conditions over land will improve the accuracy of seasonal prediction of water cycles over land. Usage of better LSMs in GCMs and satellite derived surface conditions, such as soil moisture, may contribute for the improvement, however, how land surface conditions affect the prediction of hydrological cycles over land should also be investigated. For that purpose, the GLACE (Global Land-Atmosphere Coupling Experiment) under GEWEX/GLASS is planned.

Statistical approaches may be useful to give predictions, as well. Even though that could be only for ad-hoc applications, new insights in tele-connection of hydro-climatology may be found from the sensitivity tests of the statistical models.

The prediction of climate change, particularly global warming itself, is attracting a lot of attention. Since most of the reliable predictions of hydrological cycles in the future are estimated by GCM simulations, the way to interpret or transform the GCM estimates to water resources assessment is another key issue. Down-scaling the coarse spatial (and temporal) resolution of GCM estimates is one of the scientific problems, but there are many technical issues, such as bias reduction of GCM estimates and the reproductivity of statistical characteristics particularly on precipitation simulated in GCMs. How to extract robust information from GCM predictions on water cycles in the future should be given more attention.

As for the human dimension side, how to draw the prospect on water demand, withdrawals, and consumption is basically needed. They are generally estimated with the outlook of the future growth of population, economy, etc., but it is very difficult to fully include the impact of innovation in technology and change in policy. Therefore, somehow scenarios should be set to assess the future demand in water resources. Even though socio-economic scenarios are used in the projection of future climate in IPCC, most of them do not include the feed back mechanism of how water shortages suppress various social activities and economical growth.

There is no guarantee that complex models will provide better results, but interactive modeling on water consists of natural climate system, hydrological cycles, anthropogenic activities such as irrigation, regulation by reservoirs, landuse change, etc., may provide new insights in the global water resources assessments. Virtual water trade, how much domestic water resources are saved by international trade of products, may be incorporated with the modeling of "real water."

An improved understanding of the operation of hydrological processes at the local scale is also required; *e.g.*, in order to understand how land-use change and management affect water resources and also to downscale from global models.

Regional studies should also be important for policy making, and it is recommended that information on water demand and water management should also be collected when hydro-climatic regional studies will be taken. Based on the regional studies, science-based decision-making tools should be developed, which can cooperate with the uncertainties included in each component of prediction.

Global studies are also expected to provide useful information for regional studies. In practice, there are many places in the world where global datasets of precipitation, landuse, soil types, etc., are the only sources available for hydro-climatological studies. Therefore, the applicabilities of this information on integrated water resources management on local/regional scales will be examined under the Project for

Prediction of Ungauged Basins (PUB) of IAHS (International Association of Hydrological Sciences).

5.3.3 IAHS Hydrology 2020 Working Group

The Hydrology 2020 group

(http://hydro.iis.u-tokyo.ac.jp/H2020/)

was assembled by IAHS to assess how the discipline of hydrology should evolve to meet the world water challenges that are expected to prevail by 2020. The 2020 Group would follow the pattern established by the Hydrology 2000 Working Group which was set up in Hamburg in 1983 and reported in Vancouver in 1987 [*Kundzewicz et al.*, 1987].

Nine commissions under IAHS nominated the names of two young hydrologists who are this year aged 35 or less, considering region and gender in the choice, and the president of IAHS determined that one of these two be the member of the group. WMO and UNESCO nominated one each, and the chair of the group was separately nominated by the president of IAHS, as well. The group was authorized at the plenary meeting of IAHS general assembly at Maastricht in 2001 and expected to report their vision of hydrology in 2020 at the IAHS general assembly meeting in 2005 with articles on an IAHS publication ("redbook").

The group held face-to-face meetings twice in 2001, defined their mission, and drew their road map to synthesize their visions. At first, they looked back 20 years ago, reviewed what were the most relevant research achievements in those days, and defined the changes since then in each subdiscipline in hydrology. Then, they examined the 20 years ahead with questions like:

- what will be the hydrology in 2020 under "business as usual" scenario?

- what shall be the hydrology in 2020?

- what should be done now to achieve the preferable situation of hydrology in 2020?

The group prepared their intermediate report to stimulate discussions with the general audience of IAHS/IUGG in their open workshop in Sapporo in July 2003.

The intermediate report consists of:

- uniqueness of hydrology

- advances in technology

- needs of society

- critical bottlenecks

The drafts of the latter two chapters are included in this report.

The most critical "bottlenecks," or barriers to the advancement of hydrological science, as well as some solutions necessary to overcome the obstacles, are described below. Organizational and technical bottlenecks are presented in the recommendation section.

5.3.4 Scientific Bottlenecks

The Hydrology 2020 group has recognized many fundamental bottlenecks within hydrological sciences that must be reconciled in order for hydrology to meet the water scarcity and pollution challenges of the next few decades. These scientific bottlenecks include persistent obstacles that cut across many disciplines, such as data integration, scaling, and numerical representation of complex processes. However, some of these bottlenecks are associated with the lack of understanding of fundamental processes, systems, or cycles that are most germane to hydrologists, such as the water cycle and the vadose zone. For some problems, the theory and approaches within individual subdisciplines are quite advanced, but the links to neighboring disciplines are often not well established. The individual scientific bottlenecks are described below.

Incomplete Understanding of Hydrological Processes

Deficiencies in our understanding of many of the hydrological processes and systems greatly handicap our efforts to guide water resource planning, to predict hydrologic extremes, and to predict contaminant migration.

For example, as crucial as the water cycle is to human and ecosystem existence, there are still many gaps in the understanding of the individual components that comprise the water cycle, as well as the interactions between these components. Additionally, the vadose zone, which supports our agriculture, serves as the repository for most of our municipal, industrial and government wastes and contaminants, and supports our infrastructure, plays a crucial role in both the water cycle and in contaminant hydrology. In spite of its importance, the understanding and prediction of flow and transport of water and contaminants through the vadose zone and across the upper (soil-air) and lower (groundwater) boundaries are inadequate.

Improved understanding of the hydrological processes and system components can be obtained through nested and coupled experiments replete with field measurements using several techniques, remote sensing measurements, process investigation, and modeling. Improved observations of water vapor, clouds, precipitation, evaporation, surface and subsurface runoff, groundwater, soil moisture, snow and ice using new technologies and networks collected over various spatial and temporal scales are needed to improve our understanding of the water cycle. These measurements must have enhanced resolution, increased accuracy, and be maintained for long time periods relative to the currently available data acquisition schemes. These measurements must be coupled with process studies of the individual components as well as the linkages between components over various spatiotemporal scales. The measurements and process studies must be incorporated into improved models (see below). Establishing coordinated and linked field observations with remote sensing data and modeling efforts over a variety of spatial and temporal scales is necessary for an improved understanding and prediction of water cycle, hydrological extremes, and contaminant infiltration.

Successful implementation of the nested and coupled field experiments will require advances in our approaches to modeling, data integration, and scaling. These topics are also current scientific bottlenecks, as described below.

Modeling

Many numerical models are based on hydrological principles; these models range from flow and transport in the groundwater to global circulation models. In most cases, a combination of both incomplete theory and data exist. Because of these inadequacies, many different modeling approaches have been developed, ranging from deterministic to stochastic and from data-driven to physics-based. Modeling should continue to be a research topic, with more emphasis on uncertainty assessment, improved parameterization and data assimilation approaches, standardized validation procedures, improved frameworks for incorporating indirect data (tracers, geophysics, remote sensing, etc.) and improved numerical representation of coupled processes (such as hydrogeological-biogeochemical) and systems (such as land and atmosphere).

Data Integration/Calibration

Various data sets, such as tracers, ground-based geophysics, and remote Sensing, are becoming increasingly available for use in hydrological studies. More research is needed to fully develop the potential that these tools have for assisting with hydrological problems. Many current obstacles are associated with improving the accuracy and resolution of the geophysical and remote sensing measurements. With both data types, a better understanding of data integration approaches is needed to enable routine calibration (often using data sets that are collected at different spatial scales) or to facilitate a comprehensive interpretation. With both ground-based geophysics and remote sensing data sets, we should strive to move beyond site-specific inference based on spatial patterns toward an improved understanding of the physics, so that the data can be used to more generally and quantitatively estimate hydrological-climatological parameters of interest.

Scale problems

Issues related to scale are persistent across all aspects of hydrological sciences as well as in many other disciplines. Other topics requiring additional research include investigation of dominant processes and interactions between processes that occur at different spatial scales, and reconciling the different spatial scales associated with measurements, physical processes and numerical models.

5.4 Recommendations

5.4.1 Organizational Issue

Water issues are complex, as water acts as a link between various water uses, land use and ecosystems. Calls are often made for integrated water resources management strategies, such as basin-wide hydrosolaridarity, which focuses on water allocation principles based on equity and efficiency. However, such concepts are still poorly defined and even more poorly understood. It is clear that effective policies are needed to strike compromises among competing water uses, to develop sustainable water resource plans, to alleviate and mitigate pollution, and to allay water-related disasters. These policies must be driven by scientific knowledge and scientifically based recommendations, which stem from appropriately directed hydrological research and adequate funding.

Although water quality and resource management cut across political and national boundaries, no strong and well-funded intergovernmental global hydrological organization exists that can coordinate and fund research and operational efforts within the broad range of hydrological sciences. A global hydrological intergovernmental organization is needed to serve as the authoritative scientific voice of hydrology and also to organize research efforts toward world water problems. It could have a role in facilitating integrated approaches to water development and management and offer a capacity to provide advisory services and implement and strengthen technical cooperation and investment projects targeting critical areas of water resource management. Scientific results must be translated into action-oriented recommendations so that they can be used in national and international policy evaluation, formulation, and planning. These recommendations should be formulated in terms that are clear, specific and realistic. With substantial funds and commitment, this organization could be developed anew, or the hydrology and water management sections of WMO/UNSECO and other UN agencies could be combined and expanded to meet the organizational needs of the hydrological sciences. The primary responsibilities of this global organization would include:

Water policy : The organization would both coordinate and contribute to water policy activities, serving a the central spokes-organization for global hydrology and water management;

Coordinated Research Management : A central focus of the organization would be to develop, fund and coordinate long-term research programs. Included in this task is the coordination of research programs that are developed by individual countries and organizations in order to most effectively tackle the existing scientific challenges.

Testing Centers : The organization would be responsible for establishing linked testing centers/areas that scientists can use to test models and approaches, to share data/instrumentation, and to train students or representatives from developing countries. The centers should work together to establish acceptable data standards, formats and calibration approaches and to coordinate long-term data acquisition, archiving, and dissemination. These centers are currently critically needed, yet are too costly to develop by individual researchers or small groups of researchers.

Education/Outreach : The organization should establish standards and advocate public education in hydrological sciences. The organization should oversee outreach efforts to engage young and bright scientists into the discipline, and to make formal connections between the many sub-disciplines that interact with hydrology.

Public Awareness : An important component of the organization will be to raise public awareness about the impending water crisis in a manner similar to how awareness was raised about climate change and ozone depletion issues. Greater awareness translates into more political support and more funding geared toward hydrological sciences.

Technology Transfer and Capacity Building : The global hydrological organization would strive to strike a balance between supporting fundamental research and devoting resources toward capacity building and toward finding practical solutions to hydrological and water management problems in developing countries. In many locations, there is a current disconnect between the state-of-the-art in hydrology (represented by complex research advances) and the state-of-the-practice (the tendency to implement the developed approaches in the field). While many hydrologists in developed nations focus on issues such as resolution, uncertainty, and accuracy of advanced prediction/estimation approaches, those in less developed nations are enthusiastic about, for example, the development and dissemination of inexpensive treadle pumps that can deliver irrigation water to their crops using human (cycling) power. Thus, in addition to support of fundamental research (which will be described below), there also needs to be support and guidance for well-trained hydrologists and water managers who can focus on solving practical solutions, often in the face of incomplete fundamental theory, using currently available approaches and instrumentation.

There have been unsuccessful attempts to develop such a global organization in the past. The attempts met with resistance, not only because of protectionism among existing organizations, but because of the lack of willingness to support additional intergovernmental organizations among national funding agencies. Water issues are far too often considered to be part of national security with a strong unwillingness to open up for international cooperation. However, the current situation, typified by a lack of coordination and the duplication of efforts, is not an effective way to handle the challenges that we face. Water quality and quantity trends are not auspicious, and no clear solutions are on the horizon. We urge the hydrological sciences and water policy communities to reconsider the issue of developing a global hydrological intergovernmental organization that would serve as the authoritative scientific voice of hydrology and to organize research efforts toward world water problems.

5.4.2 Practical/Technological Issues

There are several issues where funding or commitment is the major impediment to progress rather than development of new theory. These practical/technological bottlenecks are briefly described below.

Inexpensive access to data : Many of these obstacles have to do with acquisition of and access to data. In order to better assess and manage the world's water supplies and prevent hydrological catastrophes, free access to current hydrological data is needed, concomitant with long-term commitment to establishing and maintaining monitoring networks.

New Sensors : New sensors are needed to obtain improved data sets. Although microsensors that are cheap, small, automated, smart, injectable and innocuous are currently being developed in engineering and biomedical fields, development of such sensors for hydrological sciences has been sluggish. On the other end of the spatial scale, satellite sensors, which have hydrological applications as the primary goal, should be developed. These sensors should be improved to offer higher spatial resolution over more specific spectral bands, and should have faster return periods.

Water Database development and management : Another practical bottleneck within the hydrological community is the lack of a worldwide water resources database. Water resources cannot be managed unless we know where they are, in what quantity and quality, and how variable they are likely to be in the future. Development and maintenance of a database using high quality data is crucial for assessing global water resources and proper planning for their conservation. The Global Runoff Data Center is currently establishing a data base that compiles freshwater fluxes and long term mean monthly discharges from selected monitoring stations. Databases such as this should be expanded to include complete information of all significant world aquifers that includes information about capacity, water balances, purpose, and type of runoff control and associated GIS coordinates.

Technologies for Developing Countries : Development of small-scale technologies and approaches for dealing with (often nation-specific) water supply and sanitation issues could dramatically improve conditions in many developing countries. Although development and deployment of such technologies could be performed with minimal funds, a global commitment is needed to both identify the needs and to organize the effort.

5.5 Social Benefits

5.5.1 Science as part of capacity building and governance

The lack of financial resources in the water sector is only one aspect of the problem. There are other aspects that are equally important. Referring to Chapter 18 of Agenda 21, it is stated that:

> "In creating the enabling environment"... "the role of governments includes mobilization of financial and human resources, legislation, standard setting and other regulatory functions, monitoring and assessment of the use of water and land resources and creating the opportunities for public participation."

The role of science is implicitly a part of this. To have a fundamental knowledge base is crucial. This can be to strengthen international co-operation in studying processes related to the Earth's atmosphere, hydrosphere, biosphere and geosphere as a way to strengthen our understanding of water resources and the movement of water through the systems. But it is also essential to develop viable and scientifically based solutions to current and potential future problems. Although understanding the interconnections between the different systems of the Earth is recognized as being essential, it remains a sad fact that they are often dealt with independently of each other, not least in the scientific community. Promoting exchange of knowledge and ideas through improved communication among experts belonging to the various disciplines concerned will be a continuous challenge ahead. Scientists must dare to leave their compartmentalized thinking and policymakers must encourage and promote such dialogues.

Science also has a role in the development of new technologies, such as remote sensing and Geographical Information Systems (GIS) and demonstrate the applications of such technologies to managers. They can offer the potential of increasing the capacity for monitoring of the relevant elements at a moderate price. Information must then be made available. It must be recognized, however, that such technologies are often associated with initial training and technology needs, especially in developing countries. Many international organizations will have to play a more pro-active and supporting role in the endeavor to promote the use of such technologies, and support the necessary technological exchange and training associated with them. Strong partnerships between academic institutions in developed and developing countries are an important aspect, and should be promoted by governments and international organizations.

5.5.2 Bridging the communication gap

Scientific projects need increasingly try to respond to, or at least be linked to, issues relevant to socio-economic development. This is important, as there are signs of declining respect for science and scientists over recent years from policy and decision-makers. More than two thousand years ago, Socrates stated that "There is only one evil for humanity – ignorance." Political ignorance is indeed dangerous, but probably some of this declining respect could partly be attributed to the inability of scientists to present relevant information and guidance in response to emerging issues, as well as in a form appropriate for policy and decision making in broader terms. To provide "yes or no" answers is difficult, but the precautionary principle, one of the Rio Principles, provide a tool where the "best guess" approach can be applicable.

Hydrological sciences are no exception, and it can even be argued that other scientific disciplines are more active in providing policy guidance in this field, which is sometimes a matter of serious concern. A problem, which often arises, is that policy-makers within water management are frequently asking long-term questions, while many scientific programmes are restrained by short-term funding. Despite such limitations, scientist involved in such programmes can intensify their efforts to enhance the visibility and applicability of scientific results by using pro-active ways of communicating them and by better addressing issues and provide guidance in response to specific societal needs.

In the international scientific debate (on water as well as on other issues) there is a tendency among scientists (including those dealing with water) to give to much weight to their own area of expertise. It is assumed that if we only are able to do this or that research-project the world would be in better shape. Still, the improvements are forthcoming. While many scientists think that good research results almost automatically permeate into policy this is seldom (if ever!) the case. Rather, if we are to understand why certain issues are highlighted and others are not we should analyze the actual policy process and what is decisive in that process. Communication strategies need to be formulated that would allow scientists to identify better communications routes to "market" their results. Otherwise our understanding will remain limited with regards to why certain aspects or research, which are highlighted by international water scientists, are incorporated in the policy of states and international organizations and why others are not.

It seems imperative for the water specialists of the world to incorporate issues such political feasibility, ideology and cultural aspects into their analysis. The politics of water is not only about politicians, but also deals with how water specialists interact with government agencies, international organizations and NGOs.

5.5.3 Develop a scientifically based policy language

There are certain concepts that can be developed to respond to the need of policymakers while being firmly based on scientific understanding and knowledge. Such innovative concepts can help to progress thinking and act as catalyst for more progressive policy making – responding to the needs of people while recognizing the boundaries set by the hydrological situation at each given point. Some examples are presented here:

The concept of "sanctioned discourse" essentially refers a normative paradigm within which certain hypothesis might be raised why others may not. Thus, the sanctioned discourse sets the "boundaries" for what is politically feasible to do. For example, if water scientists in a water scarce region have agreed that the rational way to allocate the water would be to decrease the amounts of water being allocated to irrigated agriculture. Still, this is not being done, in spite of scientists preaching it. Why? It is argued that this is so since politicians have to act within a limited arena in which they, in this case might be heavily influenced by a strong farming lobby that pressure them to maintain existing allocations to agriculture. Thus, the sanctioned discourse sets the boundaries for action and thereby deter the politicians to act on the advise of the scientists.

The Stockholm International Water Institute has developed a concept called "Hydrosolidarity." Hydrosolidarity could be developed into a commonly accepted framework and thus provide guidelines for stakeholder interactions and problem solving in a river basin. Rather than offering a universally agreed set of principles, the hydrosolidarity concept should be based upon what could be called commonly accepted thinking in relation to basic needs of different stakeholders within a basin, where the joint water resource has to be shared between both societal activities and vital ecosystem functions. Besides that, it should offer a set of recommendations linked not only to principles and norms but also to legislative and enforcement aspects of water management – recommendations

that could be used as a fundamental base for stakeholder discussions, interactions and conflict reconciliation.

5.5.4 Recommendations for future actions – filling the gaps

The future will present us with many challenges that we, the next generation of scientist and policy makers, must handle. Some of them are already known, others will emerge. What will be the issues 1, 5, 10 years into the future?

Strong leadership is essential, based on both a positive vision on what can be achieved, but also equally important on long term persistence, not least within science. This leadership must be well informed. Only if the scientific communities work together with policy and decision-makers can water issues be successfully addressed. An integrated approach to freshwater management is the way forward as it offers the means of reconciling competing demands with dwindling supplies and a framework in which hard choices, and priorities, can be made, and effective responsive action taken.

We would like, in this paper, to present some points that we feel need to be considered to facilitate further progress:

- Knowledge and understanding of water issues is essential and must be further encouraged. National and international information networks, using modern technologies, must be strengthened. This is especially crucial in many developing countries, where international assistance needs to be enhanced;

- Efforts should be directed at facilitating the international exchange of hydrological and related data and products, so that global studies of freshwater resources and its links to socio-economic and environmental issues can be conducted and useful results produced of benefit to humankind;

- Scientific results must be translated into action-oriented recommendations so that they can be used in national and international policy evaluation, formulation, and planning. These should be formulated as clear, specific and measurable goals. This is crucial if the scientific community is to strengthen their credibility and further enhance the possibility of receiving financial support for what are sometimes costly long term projects;

- The scientific community has to find innovative ways of improving collaboration among scientific disciplines, and with new important actors such as the private sector;

- International, regional and national organizations should be encouraged to find innovative ways to strengthen cooperation among countries which share river basins or aquifers, in particular through bilateral or other intergovernmental mechanisms. This includes the development

51

of agreements to share data and other information, and joint scientific studies of shared resources;

- International organizations, including within the scientific community, must, within their mandate, continue to be well focussed and action oriented and, in particular, enhance their efforts to strengthen the capacity of many developing countries to deal with complex environmental and socio-economic issues.

5.6 Conclusions

According to the paradigm shift of research in natural sciences, it is the era for geosciences to study the real situation of the Earth, including the various impact of anthropogenic activities. Water cycles are among the most exposed nature and are vulnerable to human impacts. Therefore, hydrological science should deal with water cycles on the Earth, their impact on human society, and the anthropogenic impact on the water cycles on the Earth. Regional characteristics and historical circumstances should be considered particularly in the field of hydrological sciences.

The ultimate goal of the hydrological science is to increase the ability of monitoring, modeling, and predicting the quantity and quality of water and their input/output to the system. The outcomes of hydrological researches should have the capability to be used as a tool to know, to estimate, to understand, and to assess the water cycles on the Earth on various temporal and spatial scales by other scientific disciplines, the general public, and even by decision makers.

Scientifically excellent research and socially relevant research are not necessarily exclusive, and recent situations anticipating world water crisis gives/widens a lot of opportunities that excellent scientific research can contribute to solve practical problems. International and interdisciplinary frame works are also prepared to promote the hydrological sciences. More collaborations within hydrological sciences and with other disciplines is crucial, and formulating a system of hydrological sciences philosophically and also institutionally should be required in coming years.

IAHS Hydrology 2020 Working Group

Name	Country	lead authorship
Jeanna Balonishnikova	Russia	
	Ukraine	
Wolfgang Diernhofer	Austria	
Pierre Etchevers	France	
Stewart Franks	Australia	
Guobin Fu	China	
Kate Heal	UK	
Susan Hubbard	USA	bottle necks
Harouna Karambiri	Burkina Faso	
Johan Kuylenstierna	Sweden	social benefits
Taikan Oki	Japan	editor
Stephan Uhlenbrook	Germany	
Caterina Valeo	Canada	

Bibliography

Bras, R., and P. S. Eagleson, Hydrology, the forgotten earth science, *EOS Trans. Amer. Geophys. U.*, *68*, (16), 1987.

Buras, Nathan, *Reflections on hydrology' science and practice*, American Geophysical Union, 1997.

Crutzen, P. J., Geology of mankind - the anthropocene, *Nature*, *415*, 23, 2002.

Kayane, Isamu, Current status of researches on hydrological cycle and water balance, In *Hydrological cycles and Water Balances*, volume 167 of *Meterological Monograph*, pages 1–20. Meteorological Society of Japan, 1989, (in Japanese).

Klemeŝ, V., A hydrological perspective, *J. Hydrol.*, *100*, 3–28, 1988.

Kundzewicz, Zbigniew W., Lars Gottschalk, and Bruce Webb, editors, *Hydrology 2000*, Number 171 in IAHS Publication. IAHS, 1987.

NRC, , *Opportunities in the hydrologic sciences*, National Academy Press, 1991, Committee on Opportunities in the hydrologic Sciences Water Science and Technology Board Commission on Geosciences, Environment and Resources National Research Council.

UN, , UNDP, UNEP, FAO, UNESCO, WMO, World Bank, WHO, UNIDO, and SEI, *Comprehensive Assessment of the Freshwater Resources of the World*, page pp.33, World Meteorological Organization, 1997.

UNESCO, , and WMO, *International Glossary of Hydrology*, page pp.413, World Meteorological Organization, 1992, second edition.

Chapter 6

International Association of Geodesy

L. Sánchez

6.1   Introduction

Geodesy is classically defined as "the science of the measurement and mapping of the Earth's surface" (Helmert, 1880). This definition includes "the determination of the Earth's external gravity field," and it is complemented by the new concept of the temporal variations of both surface and gravity field of the Earth, i. e. "four-dimensional Geodesy" (Torge, 2001). These activities are also extended to the study of other celestial bodies (e. g., planets and natural satellites).

The study of the gravity field leads to the determination of the geoid. It is the main equipotential surface of the Earth's gravity field and would coincide with the ocean surface without ocean currents and other disturbances, i. e., with the hypothetical sea surface. It is the natural reference to define the vertical position: the height. It is also defined as the physical (or gravimetric) reference system. With its determination the representation of 70% of the Earth's surface (the oceans) is achieved.

The representation of the Earth's solid surface is accomplished point-wise by the coordinates of control points. These coordinates can be expressed as a three-dimensional position in Euclidean space [X, Y, Z], or as a combination of curvilinear surface coordinates [Latitude (φ), Longitude (λ)] with the vertical position [height (h)]. Since the currently available techniques to determine coordinates on the Earth's surface involve the observation of celestial bodies (natural and artificial satellites, extragalactic radio sources, etc.), the coordinates are defined according to conventional reference systems, which are implemented by two levels: a celestial reference system to describe the Earth's orientation in space and, consequently, its relative position and orientation with respect to the observed celestial bodies, and a terrestrial reference system to describe the position of the detailed surface with respect to the Earth's body. The connection between celestial and terrestrial reference systems is given by the Earth rotation parameters (precession, nutation, polar motion, siderial time and length of day). The two reference systems are also known as geometric reference systems.

The coordinates on the Earth's surface change in time due to geodynamic processes (plate tectonics, earthquakes, fluid displacements, etc.), Earth rotation and gravity field variations. Therefore, the position of a point must be complemented by a reference date, the origin of the fourth coordinate: the time.

Geodesy performs its tasks through collection, analysis and modelling of observation data. Availability and quality are strongly related to the technological advances in measurement and processing techniques. The precision level of the geodetic observations achieved today forces us to develop adequate analysis criteria and to improve theory. This forms a new basis to refine the measurement techniques. The ultimate goal is the best representation of the figure, rotation (or orientation) and gravity field of the Earth and planets, together with their temporal variations. This on going cycle leads the geodesists to explore the actions of other geosciences and to accomplish a closed feedback, the final purpose of which is the welfare of mankind.

Almost every human activity requires a graphical or digital representation of the Earth. As a consequence, an adequate reference platform (coordinate definition), primary products of geodesy, constitute the fundament to formulate and to execute a big variety of projects (planning and development, engineering, communications, navigation, natural hazard mitigation, energetic resources exploration, etc). In this way, geodesy is also engaged in facilitating and making understandable its procedures to the "every day" users, in order to intensify the reciprocal benefits.

This document outlines the most important activities and procedures, the geodesists under the umbrella of the International Association of Geodesy (IAG) wish to implement in the coming years to reach the new goals that the state-of-the-art imposes to geodesy under the global policy of achieving the maximum benefit for the scientific community and society in general.

The completion of this document would not be possible without the kind help of the members of the Planning Group of the IGGOS project (see below), who have supplied all their valuable material, and many geodesists who answered a survey questionnaire designed by the IUGG's working group Geosciences: The Future. This support is greatly appreciated.

6.2 A view of the future of research in Geodesy

The technological revolution in the second half of the twentieth century has strongly influenced the development of geodesy. In particular, the advances of artificial satellites and space-geodetic measurement techniques made it possible to conceive the Earth as an unique dynamical system. Nowadays, it is possible to acquire large observation data sets with high precision and resolution in space and time, where the influences of the physics of the Earth's interior, the interaction with Sun, Moon and planets, and the relationship between solid Earth, ice, oceans and atmosphere are identifiable. The precision of the modern observational data asks for the improvement of the theoretical models. High precision measurements and theory lead geodesy into the 1 ppb - accuracy level. Our future activities are aiming at this level, to keep it consistent over long time periods, and to make it available and useful to society.

6.2.1 Long term focus for Geodesy

To attain a better understanding of the functioning of our planet and to increase the welfare of Mankind

The shape of the Earth's surface and its changes are produced not only by the Earth's rotation and gravity field, but also by the interaction of diverse phenomena as a response to the dynamics of the Earth as an integrated system. Among these phenomena one can name the convection processes in the Earth's interior (with plate motion, crustal deformation, earthquakes, volcano eruptions), exogenic forces (such as tides, solar radiation, winds, and weather), tectonic processes (mountain building, faulting, sedimentation) and the interaction between solid Earth, ice, oceans and atmosphere (sea level rise and fall, tides, circulation and eddy motion in the oceans, ice sheet defrost). These phenomena determine the environmental conditions of the Earth. In this way, geodesy, as the geoscience in charge of the representation of the Earth's surface, provides direct measurements of these phenomena. Its procedures and products provide the strategic tools to continue, improve and preserve human life on this planet.

6.2.2 Short term focus for Geodesy

To integrate the three fundamental pillars of Geodesy - geometry and kinematics, Earth orientation and rotation, and gravity field and dynamics- to achieve the maximum benefit for the scientific community and society in general

Short term priorities

1. **Integrating different techniques, models and approaches to achieve a better consistency, long-term reliability and understanding of geodetic, geophysical, geodynamical and global change processes:** The use of space techniques in geodesy complemented by terrestrial and airborne methods allows us, firstly, to collect large data sets, which are global, i.e., their coverage over the world is homogeneous and consistent, and secondly, generate time series of measurements within reasonable intervals and with a reasonable repeat period to know in more detail the (temporal) characteristics of the Earth as an unique dynamic system. Examples of the usefulness of space geodetic techniques are: Direct and precise measurement of the continental tectonic plates movements with Very Long Baseline Interferometry (VLBI), Global Positioning System (GPS), and Satellite Laser Ranging (SLR). Quantification of small Earth crust displacements in seismic/volcano areas using GPS continuously observing stations. Monitoring of superficial deformation, displacement and stress accumulations in seismic/volcano areas using interferometric synthetic aperture radar images (mission InSAR) combined with satellite positioning. Study of the interaction between atmosphere and solid Earth by monitoring length of day and polar motion. Observation of the Earth's masses redistribution, including postglacial rebound, ocean circulation, deformation due to the mass exchange between atmosphere, oceans, ice and solid Earth, etc. with the new gravity satellite missions. Analysis of the sea level variations using satellite altimetry, including ocean circulation and its effects in global climate events, such as El Niño. Table 1 list the relevant space-geodetic techniques and their achievements. This list is far from complete, but it gives a general idea about the challenges that the geodesists have to face in the next decade(s).

2. **Consistent modelling of observable signals of the Earth's system:** Modern geodetic observing procedures are mainly based on space techniques (VLBI, satellite positioning systems, satellite altimetry, satellite gravity missions), which provide global orientation, and are affected by the same physical (e. g., atmospherical) factors. Often the processing and analysis of the observed data is carried out "independently" using different standards and models. To take maximum benefit of these data, i. e., to produce consistent information, it is necessary to use one and the same fundamental reference system, which unifies constants, standards, geometric and gravitational reference frames, models and parameters. This is the motivation for the establishment of the Integrated Global Geodetic Observing System (IGGOS). With this Project, the scientific and infrastructural ba-

Table 6.1: Space techniques in geodesy

Geodetic Area	Use	Space Techniques/Space Missions
Geometry and kinematics	Geometrical reference systems, global deformation processes, temporal variations of the Earth's surface (solid, ice and ocean areas), this includes plate tectonics, earthquakes, volcanoes, natural hazards, climate change, etc.	VLBI (Very Long Baseline Interferometry) * SLR (Satellite Laser Ranging) LLR (Lunar Laser Ranging) * DORIS (Doppler Orbitography and Radiopositioning Integrated by Satellite) * GPS (Global Positioning System) * GLONASS (Global Navigation Satellite System) * GALILEO (Global Navigation System) * T/P (TOPEX/Poseidon - Ocean Topography Experiment) * ERS-2 (European Remote Sensing Satellite - 2) * GFO (GeoSat Follow-On) * JASON (First altimetric satellite follow-on mission to TOPEX/Poseidon) * ENVISAT (Environmental Satellite) * InSAR (Interferometric Synthetic Aperture Radar) * ICESAT (Ice, Cloud, and land Elevation Satellite) * CRYOSAT (Altimetry mission to determine variations in the thickness of the Earth's continental ice sheets and marine ice cover)
Earth's rotation and orientation	Polar motion and length of day variation (oscillations of the Earth's rotation, caused by mass interchange between atmosphere, ocean, ice sheets, and solid Earth)	VLBI (Very Long Baseline Interferometry) * SLR (Satellite Laser Ranging) * LLR (Lunar Laser Ranging) * GPS (Global Positioning System) * DORIS (Doppler Orbitography and Radiopositioning Integrated by Satellite)
Earth's gravity field and dynamics	Physical reference system, mass balance, quantification of mass displacements (post glacial rebound, ocean circulation, hydrology, lithosphere)	CHAMP (Challenging Microsatellite Payload for Geophysical Research and Application) * GRACE (Gravity Recovery and Climate Experiment) * GOCE (Global Ocean and Climate Experiment) * Altimetry (T/P, ERS-2, GFO, JASON, ENVISAT)

sis for all geodetic global change research shall be provided, viewing the Earth system as a whole by including the solid Earth as well as the fluid components, the stationary and time-varying, geometry and gravity field.

3. **Combining consistently geometrical and physical references systems:** The widespread use of the space-geodetic techniques, in particular GPS, in the daily life of non-geodetic, "normal" citizens requires an interface between the developers (scientists) and the users, not only to manage the technology contained in an instrument, but also to understand the scientific background making the application of this technique useful and successful. The "mm-level" precision, at times demanded by the users, is not yet disposable. Such requirements can be achieved eventually by continuously improving the reference system realisations (materialisation). This task includes, from the geometric point of view, the densification and maintenance of the reference frame through a large and globally homogeneous set of control points with coordinates of sub-centimetre accuracy, complemented by a velocity model at sub-millimetre per year level. The velocity model should include both horizontal and vertical movements of the Earth's crust, i. e., a three-dimensional Earth's deformation model is highly required. On the other hand, from the physical point of view, the natural reference surface for the heights, the geoid, must reach the centimetre precision level, as well. Then, a consistent relationship between the two systems should be found to remove the incompatibilities of vertical positioning. More precisely, the following must be achieved: the GPS-derived heights (h) refer to the ellipsoid (a geometrical model of the Earth), the commonly used heights (H), that are consigned on maps or in geo-information systems, refer to an equipotential surface of the Earth's gravity field (the geoid, a physical model of the Earth). The connection between these two types of heights is known as geoid undulation (N); it describes the distance between geoid and ellipsoid. The practical use of GPS demands that the equation $h = H + N$ be satisfied in the 1 cm - level. This requirement includes other important tasks, which must also be solved by geodesy. Among them are: Definition of a global vertical reference system, and, since the geoid was classically defined as a surface, which coincides with sea level, determination of the offsets between each local reference system, based on a local sea level and the global one.

Short term benefits to Society

1. **Providing better information to improve the knowledge about the system Earth:** The Integrated Global Geodetic Observing System (IGGOS) should be seen as geodesy's contribution to study the Earth system composed of solid geosphere, cryosphere, hydrosphere and atmosphere. It will provide its findings to interdisciplinary research, governmental agencies and the private sector, inform and train the customers about the correct interpretation and use of the geodetic products.

2. **Promoting interdisciplinary actions to make the geoscientific information suitable to improve the life standard on Earth:** The Integrated Global Geodetic Observing System (IGGOS) should be part of the Integrated Global Observing Strategy (IGOS), sponsored by the United Nations Educational, Scientific and Cultural Organisation (UNESCO) and the International Council for Science (ICSU). It should also be a recognized partner of the other, already existing, observing systems: Global Climate Observing System (GCOS), Global Ocean Observing System (GOOS) and Global Terrestrial Observing System (GTOS). In this way, geodesy will participate more directly in providing fundamental information "... to detect, locate, understand and warn of changes in the terrestrial ecosystems".

3. **Supplying a global observing system for hazard assessment:** The observable signals of geodesy provide the basis to identify regions with high risk in natural disasters. For instance, the repeated geodetic observation of crustal deformation zones allows us to locate possible break points during earthquakes. The study of the gravity field variations in time leads to the detection of mass movements, for instance, in volcanic areas, i. e., to the prediction of possible eruptions. Satellite altimetry allows scientists to forecast El Niño phenomena, the influence of which on the global climate (and also on agriculture, agronomy, hydrology) is very strong. The sea level changes, measured also with satellite altimetry, may be a response of global warming (part of the steering effect). In this way, geodesy observes the influence of the global climate changes on the ices sheets, as well. All these geodetic activities form an accurate and reliable tool for climate prediction (short-term and long-term) as well as security warning in coastal zones. There are numerous geodetic developments and applications supporting the formulation of policies to prevent and to mitigate natural disasters. In order to achieve the necessary reliability level, the combined work of all Earth sciences is required.

4. **Providing a precise and reliable reference frame for navigation, engineering, cadastre, geo-information:** The combination of the physical and geometrical reference systems, together with the improvement of their components, support the users to reach 1 cm - level in navigation, to increase up to centimetre level the kinematic and real time positioning accuracy, and in general, to build a consistent global spatial data infrastructure under the new concept of the digital Earth.

5. **Assisting international cooperation for developing regions:** The global concept of geodesy requires a standardisation of the procedures world-wide. In this context, the technical transfer to developing countries should be done at all levels: Densification of control points, homogenisation of the observing and analysis techniques, educational strategies, international projects, joint meetings and reciprocal benefits in using geodetic information to sustainable development, including hazard mitigation.

In accomplishing the above-mentioned goals, the contribution of all geodetic disciplines is required. The field of geodesy is vast, but each of its areas, studying different phenomena and looking for the correlations to achieve the best results, observe one and the same Earth. The convergence of all these protagonists takes place under the auspicies of the International Association of Geodesy (IAG). In this framework, a general assessment of the current activities in the main structure of IAG is provided in the following paragraphs.

6.3   Assessment of current research in the major geodetic disciplines

6.3.1   Reference Frames

Major scientific goals and challenges of the next decade

1. **Determination of a reliable Earth reference frame:** As the positioning techniques may refer to different reference systems, the combination of their results may generate inconsistencies that must be resolved through the definition and realisation of a common (integrated) reference system, i.e., a reference frame.

2. **Homogenisation of the geodetic control world-wide:** The geodetic control has to be realised by a set of points (pillars or instruments) with highly accurate coordinates. Because the space-geodetic techniques (except GPS) are very expensive, the control points are, at present, localised mainly in Europe, North America, Japan and Australia, with a clear unbalance towards the northern hemisphere. To improve the observing system and its results, it is necessary to achieve a more homogeneus global distribution.

3. **Velocity field models and time dependence of the coordinates:** The position of a geodetic site is described by three cartesian coordinates $[X, Y, Z]$ and the corresponding change in time, i. e., its velocities $[V_x, V_y, V_z]$. Because the Earth is not rigid and because it is impossible to measure these displacements "everywhere," a representative model must be derived for the velocity field.

4. **Implementation of a reference infrastructure for global 1-cm navigation:** One of the most widespread applications of GPS is the precise navigation in real time. Because the propagation of microwaves through the atmosphere is not well known in real time, this kind of positioning is limited by atmospheric effects. In the specific case of navigation, the corresponding deficiencies can be overcome by using complementary terrestrial control navigation systems.

Strategies for achievement

1. **Establishment and maintainance of the (four-dimensional) geometric reference frame.** This task is primarily solved by the geometry-related IAG services, namely IERS (International Earth Rotation Service), IGS (International GPS Service), IVS (International VLBI Service) and ILRS (International Laser Ranging Service). The geometric reference frame is (must be) used for all high-accuracy applications (e. g., precise orbits for Low Earth Orbiting Satellites -LEOs-).

2. **Improvement of the global reference frame.** Design, development, and maintenance of a well- defined terrestrial IGGOS network, including all geodetic (and ancillary) techniques. The result might be called the IGGOS network.

3. **Atmospheric research using geodetic methods:** Determination of global ionosphere models based on microwave observations, high-resolution tropospheric models and deploy standardised meteorological equipment on IGGOS network.

4. **Participating in the definition and implementation of new Global Navigation Satellite Systems (GNSS).** It is IAG's declared goal to include upcoming alternative navigation systems (like GLONASS, GALILEO, ...) into its scientific work. New systems should have at least the same accuracy level as GPS.

6.3.2   Gravity Field

Major scientific goals and challenges of the next decade

1. **Implementation of a precise Earth gravity model:** Global gravity field modelling is required for large-scale problems, including the determination of dynamic satellite orbits, inertial navigation and development of geophysical and geodynamic models. This is achieved by high-degree gravity field models, which may then used to establish a global vertical datum and for the transformation of geometric GPS-heights to physical (normal or orthometric) heights.

2. **Improvement of the knowledge of the Earth's gravity field and its temporal variation:** Conventionally, the gravity field has been treated as a static field because it is dominated by the internal mass distribution of the solid Earth, which was created on time scales from millions to billions of years. With the new satellite gravity missions, complemented by airborne gravity, is possible to observe variations caused by dynamic processes (mass redistribution) that vary on time scales from hours to thousands years.

3. **Combination of terrestrial and spatial gravity field information:** This task has been studied intensively, but a complete, convincing and feasible methodology is still missing.

4. **Developing an optimal strategy for the assimilation of geodetic gravity information into global inverse geodynamical models:** This aspect has been traditionally treated in geodesy, starting from the old isostatic theory of the middle of the nineteenth century, and has proved to be important to assess modern rheological theories.

5. **Definition and establishment of a unique global vertical reference System, including the determination of a highly precise geoid (on 1- cm level):** The classical height datums are defined independently by mean sea level observations at different tide gauges over different time intervals. In this way, the combination of heights referred to different tide gauges is not reliable; discrepancies up to meter level can be expected. It is necessary to define a unique global reference level independent of spatial and temporal sea level variations. On the other hand, the comparison of GPS-derived geoid undulations ($N = h - H$) with gravimetric geoid models presents still discrepancies in the decimetre-level. It is necessary to improve the knowledge of the gravity field (gravity field satellite missions), of the solid Earth surface (spaceborne InSAR measurements), of the oceanic surface (radar altimetry), and to complete surveying of large uncovered areas, like Antarctica.

6. **Precise orbit determination, including the temporal variation of the gravity field:** This task is in essence solved for the static part of the gravity field and for the "normal" tidal variation. For temporal variations, as they will be revealed by the new space missions CHAMP, GRAVCE, GOCE, a new methodology must be developed.

7. **Identifying the relationship between the sea level change observed by satellite altimetry and global warming:** The accuracy of the sea level determination by satellite altimetry is presently approaching 10 mm. The combination of satellite altimetry and time varying gravity leads to identify the influence of mass changes

and thermal expansion on sea level variations, including secular changes.

Strategies for achievement

1. **Implementation of a collocated reference frame** with highly precise gravity instruments (absolut and relative gravimeters), selected tide gauges, and a zero-order levelling network. This task must be tackled by IGGOS.

2. **Maintenance and improvement of a unified worldwide gravity network** for monitoring gravity field time variations and providing a net of ground control points for spatial gravity surveying.

3. **Establishment and maintenance of the "official sea surface"** based on the results of all available altimeter missions. Task for the altimetry service (to be created).

4. **Establishment and maintenance of the official global model of "Earth's gravity field and geoid,"** which is consistent with all available geometric information (including sea surface) and based on results from analyses of classical geodetic satellites (Lageos, etc.), dedicated gravity missions (CHAMP, GRACE, GOCE, etc.), and oceanographic information. This is a key activity to be coordinated by IGGOS.

5. **Improvement of gravity data with high spatial resolution:** Densification of terrestrial, marine, airborne, gravimetry and satellite gravity field observations (satellite altimetry, dedicated gravity satellite missions, etc).

6. **Modelling of vertical velocities, sea level variations** and an adequate combination of terrestrial levelling techniques and the global positioning system.

7. **Implementation of a unified global height system:** as a result of these six items. They should be coordinated by IGGOS.

6.3.3 Earth Rotation and Geodynamics

Major scientific goals and challenges of the next decade

1. **Modelling of mass displacements within the system Earth from geodetic observations:** Earth rotation is governed by the three body problem Earth-Moon-Sun. The Earth has to be treated as a body of finite extensions consisting of the solid Earth (which has non-rigid properties), the oceans, and the atmosphere. Geodynamic processes play an essential role. This includes the lunar and solar gravitation and mass redistribution in the atmosphere, the hydrosphere, the solid Earth and the liquid core. The variations are secular, periodic or quasiperiodic and irregular in nature. By observing these variations, it is possible to reconstruct (to some extend) the mass (re)distribution on the system Earth.

2. **A three-dimensional Earth deformation model:** The changes in time of the coordinates on the Earth's surface are determined by repeatedly measured positions of the same points with the help of geophysical models of plate tectonics. However, the geophysical models are based on seismological and magnetic data observed in plate boundaries, showing only constant (linear) superficial (two-dimension) displacements and neglecting the inter- plate and intra-plate deformation. It is necessary to implement a global, precise, detailed, physically realistic reference model for crustal movements to include and replace the geophysical models. It should also describe the horizontal and vertical components of the deformation.

3. **Improvement of the knowledge about the remaining Earth's rotation variations:** There are still variations in the Earth's rotation rate, which are not totally understood (e. g., the driving mechanisms of the Chandler wobble and all decadal variations, to mention only two examples).

4. **Sea surface topography and sea level change:** Variations of the sea level with time are relevant in geodesy for the definition and realisation of height reference systems, especially the geoid. On the other hand, the determination and interpretation of sea level changes lead to a better understanding of the ocean-atmosphere interactions and their influence on climate. An acceleration of the global sea level rise would indicate climate changes related to the global warming.

Strategies for achievement

1. **Observation, analysis and data interpretation in tectonic regions,** post-glacial rebound and sea-level fluctuations, including the proper modelling of them and their relationship with environmental changes.

2. **Examination of the influence of the Earth's mobile constituents (fluids)** on the Earth's orientation and position in space, defined broadly in terms of length of day, polar motion, nutations, geocenter and gravity variations.

3. **Computation of time variations of angular momenta and related torques,** gravitational coefficients, and geocenter shift for all geophysical fluids based on global observational data, and/or products from state-of-the-art models.

4. **Establish ice mass balance and monitor the solid Earth** as a continuum using space techniques such as InSAR, CRYOSAT, ICESAT.

5. **Improve and maintain geophysical models** relevant for geodesy.

6. **Integration of tide gauge data with satellite altimetry** as well as oceanographic and meteorological data (sea surface temperature, salinity, current velocities, air pressure).

6.3.4 Positioning and Applications

Major scientific goals and challenges of the next decade

1. **Applications of geodetic positioning using three-dimensional geodetic networks (passive and active networks), including monitoring of deformations:** Geodetic networks consist of monumented (passive) sites or of continuously observing (active) stations. They provide the reference frame for positioning at all scales. Global networks give the global realisation of the reference systems, which form the fundamental basis for all national or continental geodetic surveys. They are the basis for geo-information systems and map series. Local networks are established for engineering and exploration projects and for geodynamic investigations.

2. **Applications of geodesy to engineering:** Positioning based on an adequate reference frame is a basic step in every engineering project. Geodesy plays an important role in the planning and implementation of engineering projects. The high precision of the geodetic methods also allows it to monitor and predict deformation and movements in engineering constructions, i. e., mines, bridges, dams, areas of geological hazard, etc.

3. **Atmospheric investigations using space geodetic techniques:** GPS radio occultation measurements are a promising tool for remote sensing of Earth's atmosphere. This technique, implemented in new satellite missions (e. g., CHAMP), offers a wide spectrum of applications in climate monitoring, weather forecast and atmospheric research.

4. **Modelling the path delay of electromagnetic waves in the atmosphere:** In practically all geodetic measurements, electromagnetic waves serve as signal carriers, this includes the methods of satellite and terrestrial geodesy as well as geodetic astronomy. When travelling through the atmosphere, the waves experience changes in velocity and in the curvature of the path.

5. **Increased accuracy of real time and absolute satellite- based positioning, including the integration of new systems for navigation and guidance of platforms:** GPS plays a dominant role in many navigation systems used today. It is beginning to play an increasingly important role in providing accurate time signals for many industries. Applications are comprised of land surveying, autonomous vehicle control, including concepts like the smart highway system, marine navigation,

air traffic control, satellite navigation, and power signal time synchronisation. By exploiting differential signals, GPS is capable of positioning with an accuracy in the order of a few millimetres.

6. **High-resolution global and regional Digital Elevation Models (DEM):** A DEM is a digital file consisting of terrain elevations for ground positions at regularly spaced horizontal grids. They are useful to model terrain and gravity data in geoid determination, surface deformation, searching energy resources, calculating the volume of proposed reservoirs, determining landslide probability, etc.

7. **The role of geodesy in the "Digital Earth" project:** Digital Earth is defined as "a virtual representation of our planet allowing it to explore and interact with the vast amounts of natural and cultural information gathered about the Earth." Digital Earth Models allow it to scientists, teachers, policy-makers, etc., to develop detailed analyses of features and phenomena as small as one meter in size. The consistent combination of all of this information is only possible if the reference frame is provided by geodesy.

Strategies for achievement

1. **Elimination or reduction of the atmospheric refraction on geodetic measurements** by improving the instrument design, observation methodology, and by the use of atmospheric models based on data collected on the Earth's surface and in space.

2. **Implementation of algorithms for accurately and reliably processing the GPS signals** for timing and navigation leads to solving problems with reflected signals (multipath), integer solutions of positioning algorithms, to determine phase ambiguities in position determination and to estimate the water content in the troposphere and the electron content in the ionosphere

3. **Improvement of the currently avaliable Digital Elevation Models (DEM),** based on digitised topographic and bathymetric information from the new space techniques. Satellite radar altimetry serves for the height determination of the Greenland and Antarctica ice sheets and provides bathymetric information, due to the correlation between the ocean surface and the ocean bottom. Space and airborne interferometric synthetic aperture radar (InSAR) has become an efficient method for the development of high resolution DEM's.

4. **Definition and maintenance of geodetic standards:** The geodetic networks serve as a three-dimensional reference for an increasing number of applications; these applications essentially require merely low level accuracy. Standards "specify the absolute or relative accuracy of a survey," and they should be derived from the objectives of the geodetic networks in terms of meeting actual needs. In this way, the users of geodetic information are in a position to plan new projects, thus reducing the actual costs and the time requirements.

5. **Strong cooperation to related surveying/navigation organisations,** such as International Federation of Surveyors (FIG), International Society of Photogrammetry and Remote Sensing (ISPRS), Institute of Navigation (ION), UNAVCO, etc., is required to achieve this goal.

6. **Participation in the implementation of initiatives related with "referencing geo-information,"** such as Digital Earth, Global Map, Global Spatial Data Infrastructure (GSDI), etc.

6.3.5 Theory

The continuous increase in number and accuracy of geodetic measurements compels the models to become more and more comprehensive in order to account for new effects, which are becoming accessible to the new measurement techniques. When a new model is developed, it may become desirable to improve the measurement procedures again testing the predictions by the theory with the reality, and the cycle begins again. This process is only possible if the theory is constantly developed in geodesy. The theory is the common basis for all geodetic areas; it is "understanding systematically the particular area of the real world, which is object of the investigation." In this way, theory does not recognise specific boundaries, and its development cannot be "planned," but it advances according to the scientific requirements. In this case, its concrete task is related to developing fundamental mathematical and physical concepts and methodologies required by geodesy. It includes the articulation of the challenges of geodesy, in terms of mathematics and physics, and common activities with scientists in other areas of science and engineering.

6.4   Recommendations

The scientific goals in geodesy and possible strategies for achievement in the next decade have been presented. There are some logistic aspects, which could contribute to the improvement of the dialog between geodesy, other Earth sciences and society in general. Therefore, some organisational recommendations can be presented below. They may be useful not only for the International Association of Geodesy (IAG), but also for the International Union of Geodesy and Geophysics (IUGG).

1. **Develop and define a consistent model for the System Earth:** IUGG should promote and coordinate the implementation of a global observing system and modelling, that includes all the disciplines of its Associations looking at the results in an integrated manner.

2. **Prediction and mitigation of natural disasters:** IUGG and its Associations should be instrumented to understand natural disasters and support the formulation of prevention and mitigation policies. In this way, an integrated vision of risk assessment supporting the action of international and interdisciplinary services in terms of recognition by international institutions (UN, UNESCO, etc.) and national institutions (governments, agencies etc.) is required.

3. **Promotion of the geosciences products:** IUGG and its Associations should increase the advertising concerning the applicability of their products. They should, e. g., explain and outline the importance of the achievements for society.

4. **Interaction with other Earth societies :** IUGG and its Associations should strive for a closer collaboration with other scientific organisations like the American Geophysical Union (AGU), European Geophysical Society (EGS), etc.

5. **Attracting young people:** IUGG and its Associations should publicize new results with the goal to attract young and bright people from mathematics, physics, astronomy etc., to its disciplines, e.g., by establishing international prizes, by fostering the creation of international doctorates in geosciences, by supporting the creation of international institutes, etc. Free and cheap publication facilities (Internet sites, journals, etc.) are also desirable.

6. **Geosciences world-wide:** IUGG and its Associations should encourage a stronger participation of geoscientists from developing countries. This can be achieved by assuring that they are able to participate in scientific collaborations (not necessarily only in meetings), like projects and programs, by making data available to them and by encouraging all nations to became members of IUGG. Another action would be to offer free membership to individual scientists.

7. **Geoinformation world-wide:** IUGG should improve the work in terms of services in its Associations and integrate them into a unique system, which is justified on a scientific ground by the IUGG structure. It should include a global policy of free and open data exchange in all fields of Earth sciences (sea level, meteorological data, geodetic data, etc).

8. **Interdisciplinary work:** IUGG should promote more interdisciplinary schools and research units at the international level, to improve mutual understanding and cooperation between its Associations.

6.5 Opportunities for interdisciplinary research

Classically, mapping was viewed as the main purpose of geodesy; it should make control surveying to provide position control for mapping. However, by establishing reference geodetic networks, geodesy became the fundamental science for disciplines related to positioning, such as urban and environmental management, engineering projects, (international and national) boundary demarcation, ecology, geography, hydrography, etc. Still other applications are found in scientific fields that have a two-way relationship with geodesy: while geodesy supplies one kind of information to them, they provide another kind of information for use in geodesy, for example: theoretical and exploration geophysics, geodynamics, space science, astronomy, oceanography, atmospheric science, geology, etc. The new techniques to observe and to analyse the Earth as one unique system make the relationship between these disciplines stronger; it is almost impossible to identify clear boundaries between their study fields. Based on these considerations and keeping in mind that IAG is one of the seven Associations of IUGG, we mention a few projects, which may be of common interest to many (not to say all) IUGG associations:

1. **Analysis and prediction of sea level variations, including ocean modelling:** IAG, IAMAS, IAPSO, IAHS, and IASPEI. Geodesy provides precise information concerning the melting of glaciers and polar icecaps, sea level change, atmospheric mass movements, ground water and soil moisture.

2. **Precise modelling of crustal deformation, including plate tectonics, volcanoes and earthquakes:** IAG, IAVCEI, and IASPEI. Geodesy precisely monitors crustal deformation (secular and episodic, extended and local), as well as gravity and Earth's rotation changes due to mass displacements.

3. **Precise modelling of the atmosphere (ionosphere and troposphere):** IAG, IAGA, and IAMAS. Geodesy provides information concerning TEC (Total Electron Content), pressure and humidity from atmospheric sounding using GPS. It might also contribute information on short-term variations.

4. **Inverse modelling of the continental lithosphere and upper mantle:** IAG, IAGA, IAVCEI, and IASPEI. Geodesy can model spatial and temporal variations of

the Earth's gravity field due to internal mass imbalances and quantify horizontal and vertical crustal movements.

5. **Observing and understanding the global climate changes:** IAG, IAMAS, IAPSO, IAGA, IAVCEI, IAHS, and IASPEI. Geodesy provides precise information about variation (spatial, secular and temporal) of the sea level and Earth's rotation rate changes due to geophysical processes, including mass redistribution effects in the atmosphere, hydrosphere, and solid Earth.

6. **Implementation of a global Earth management system:** IAG, IAMAS, IAPSO, IAGA, IAVCEI, IAHS, and IASPEI, coordinated by IUGG. It should include observation, analysis, modelling and an adequate digital representation of the Earth's dynamics as a function of space and time. This item could be seen as a combination of the four previous topics.

7. **Hazard prediction and mitigation:** IAG, IAMAS, IAPSO, IAGA, IAVCEI, IAHS, and IASPEI, coordinated by IUGG. As a continuation (or direct application) of the previous item, it should lead to develop global, regional, national and local prediction systems to prevent natural disasters (such as landslides, inundation, volcano eruptions, earthquakes, hurricanes, cyclones, etc).

These aspects do not exclude other common activities which are being developed with organisations other than IUGG, such as International Astronomical Union (IAU), International Union of Geological Sciences (IUGS), American Geophysical Union (AGU), European Geosciences Union (EGU), and some global programs of the International Council for Science (ICSU), for instance: Scientific Committee on Antarctic Research (SCAR), Program on Capacity Building in Science (PCBS), Committee in Data for Science and Technology (CODATA), Scientific Committee on Problems on the Environment (SCOPE), Scientific Committee for the International Geosphere-Biosphere Program (SC-IGBP), Special Committee for the International Decade for Natural Disaster Reduction (SC-IDNDR), Special Committee on Oceanic Research (SCOR), Science Committee on Solar-Terrestrial Physics (SCOSTEP), Committee on Space Research (COSPAR), Committee on Frequency Allocations for Radio Astronomy and Space Sciences (IUCAF), Scientific Committee on the Lithosphere (SCL), etc.

6.6 Conclusions

The presented reflections show a global assessment of the current and coming activities in geodesy. They should not be seen as isolated tasks; on the contrary, they are active components of the required analysis to better understand our planet. Geodesy is one of the oldest Earth sciences. Its challenges

and attainments share the same objective with the other geosciences: "to improve the life standards of mankind." From the sole geodetic point of view, in spite of the presented future perspective, these activities are just the improved continuation of an interminable process that began more than 2600 years ago with the first questions about the actual figure of the Earth. Now, our responsibility is keeping this process moving to inherit our best efforts to the coming generations.

Bibliography

Beutler, G. On behalf of the IAG Committee for the realisation of the new IAG structure. September 2002. http://www.gfy.ku.dk/ iag/newstructure/new_iag_s_02.htm

Beutler, G. Drewes, H. Reigber, Ch., Rummel, R. (2003): Proposal to establish the Integrated Global Geodetic Observing System (IGGOS). IGGOS Planning Group.

Dickey, J. (2001): Interdisciplinary space geodesy: Challenges in the new millennium. In: Vistas for Geodesy in the New Millennium. IAG Symposia, Vol. 125. pp. 590-594, Springer Verlag, Berlin, Heidelberg.

Helmert, F. R. (1880): Die mathematischen und physikalischen Theorien der Hoeheren Geodaesie. Teubner, Leipzig (reprint Minerva GmbH, Frankfurt a. M. 1961).

Rummel, R. (2001): Space geodesy and Earth sciences. In: Vistas for Geodesy in the New Millennium. IAG Symposia, Vol. 125, pp. 584-589, Springer Verlag, Berlin, Heidelberg.

Rummel, R., Drewes, H., Beutler, G. (2001): Integrated Global Geodetic Observing system, (IGGOS): a candidate IAG project. In: Vistas for Geodesy in the New Millennium. IAG Symposia, Vol. 125, pp. 609-614, Springer Verlag, Berlin, Heidelberg.

Sanso, F. (2001): The greeing of geodetic theory. In: Vistas for Geodesy in the New Millennium. IAG Symposia, Vol. 125, pp. 595-601, Springer Verlag, Berlin, Heidelberg.

Senatskommission fuer Geowissenchaftliche Gemeinschaftsforschung der Deutschen Froschungsgemeinschaft (1999). Geotechnologien: Das "System Erde": Vom Prozessverstaendnis zum Erdmanagment. GFZ, Potsdam.

Torge, W. (2001): Geodesy. 3rd Edition. De Gruyter, Berlin, New York.

Vanicek P., Krakiwski E. (1986): Geodesy: The concepts. North Holland, Amsterdam.

Chapter 7

International Association of Meteorological and Atmospheric Sciences

S. Adlen

7.1 Introduction

The Meteorological and Atmospheric sciences seek to understand physical processes in the atmosphere, including radiation, the physics of clouds, atmospheric electricity, boundary layer processes and atmospheric dynamics and chemistry. Despite the huge complexity of the problem, the atmospheric sciences have developed an impressive capability over the past century to help society anticipate atmospheric events. Progress is accelerating today as improved observational and remote sensing capabilities provide more accurate resolution of atmospheric processes, and enhanced physical understanding, new modeling strategies, and powerful computers combine with these measurements to result in continually improving atmospheric simulations and predictions.

The rate of progress in such a diverse field as the atmospheric and meteorological sciences is, however, dependent upon several factors. Co-operation, collaboration and dissemination of data across the full breadth of the field are essential, along with a flexible road map to provide focus and direction for research. The IUGG general assembly provides the perfect platform for reflection upon the current status of research and a re-assessment of the direction in which research in the atmospheric sciences should be progressing.

Many reports (e.g., *IPCC* [2001]) have been produced, assessing research in the atmospheric and climate sciences. This document differs by giving a young scientist's perspective on future research. Aims for the future in the atmospheric sciences are presented over long term timescales (> 50 years) along with more immediate research priorities to be pursued over the next decade.

The report emphasizes atmospheric research, but also touches upon the societal benefits of atmospheric science. Goals for the future are presented, the current direction of research in the atmospheric sciences summarised and several recommendations are suggested with the aim of moving toward a vision of the future.

Views presented within this report are those of the author alone and are intended only as the basis for discussion. The author would, however, like to thank all those who aided in the development of this document, particularly Dr. D. J. Frame, Professor F. W. Taylor, Dr. W. A. Norton and Dr. J-N Thepaut.

7.2 Long Term Goals

The relevance of science is often best measured by its value to the end user. In the case of atmospheric science, the entire population that breathes the atmosphere and experiences the variability of the Earth's weather systems could be considered an end user. If one accepts both that the human race is worth preserving and that prior knowledge of events improves "quality of life," then the value of atmospheric research is unquestionable. It is in the privileged position of being both critical to our future survival and a central feature in other "comfort factors" associated with everyday life.

To date, atmospheric research has already provided a number of ways of improving both the quality and sustainability of life. In assessing the causes of and probable future trends in climate change and in the forecasting of high impact weather, the probability of our survival is constantly being improved . At the current time, there is also a growing market in which atmospheric science benefits more specific end users, e.g., through energy, health and agriculture. It is through these benefits to society that atmospheric research finds its "raison de etre" beyond that of an extremely complex intellectual challenge.

The ultimate scientific aim of atmospheric research must be an atmospheric (and coupled Earth system) model with an infinite temporal and spatial resolution and certainty in prediction. This ultimate goal is obviously eventually limited by the underlying principles of both Quantum Mechanics and Chaos. Thankfully, current limitations in a progression toward this ultimate limit are found in difficult, but surmountable, problems, such as computing power, observational resolution and modeling inefficiencies. These limitations result in the fact that an effectively continuous problem in space and time must be represented on an unnecessarily coarse grid with an incomplete description of the physical processes describing the situation.

Techniques developed in the field have already made enormous progress in characterising and reducing errors inherent in the process of modeling the atmosphere and its evolution. However, whilst knowledge of the atmosphere and climate has improved dramatically over recent years, we are still not in the position to predict with sufficient certainty what the future holds or to exert any controlled influence over how things might change. This applies with regard to both long term (decadal to millennial) climate prediction and short term (hourly to seasonal) weather forecasting.

Climate change is arguably the greatest challenge facing mankind in the coming decades. The primary goal of atmospheric research must hence be to get to the point where we can convince all policy makers that mechanisms of climate change are adequately understood in order that detrimental anthropogenic effects on our climate can be nulled before the onset of an irrevocable dynamic. Progress in atmospheric research is, however, accelerating rapidly and goals for the next 50 years should stretch both our intellectual and engineering capabilities. A realistic but suitably challenging goal for the next 50 years is:

"To develop and integrate models of all processes (at all scales) occurring both in the atmosphere and at its boundaries, resulting in the capacity to predict events (on scales from hours to millennia) with sufficient accuracy so as to be able to understand their occurrence and minimise their effect where possible."

As previously stated, from the perspective of self-preservation, the most important role of the atmospheric sciences is the part they play in the perpetuation of the human race and in the environment in which we reside. Our improved knowledge of atmospheric processes and climate change should then be used in any possible way to enhance the probability of the continuation of the human race. The following two long term projects, utilising enhanced knowledge in the atmospheric sciences could reasonably be pursued.

- Forcing of the Atmosphere to Maintain a Habitable Environment

  If and when atmospheric processes, climate and climate change are understood sufficiently, it might be possible to intentionally influence the climate in order to sustain the Earth, improve our economies, mitigate any adverse changes in climate or null the onset of any high impact weather events.

- Location and Development of other Habitable Environments

  There is a strong possibility that anthropogenic influences upon our planet are spiraling out of control. Consideration should be put toward finding and developing other potential habitats in which the human race might survive. These habitats might include additional planetary systems

The benefits of pursuing the goals and projects outlined in this section are beyond doubt. However, all of the above long term undertakings require research over more immediate timescales in order to be able to progress and will represent the culmination of both atmospheric research and research in other related disciplines. Improvements in observations, computing and modeling will all be required in order to achieve these goals. Some more immediate priorities are outlined in the following section.

7.3 Short Term Priorities

On intermediate timescales, the atmospheric sciences must be looking to improve upon the current understanding of atmospheric processes in order to be in a position to implement and satisfactorily conclude the projects of section 7.2 if required. Recent advances in computing power are allowing unprecedented opportunities as regards atmospheric modeling. The efficient exploitation of these opportunities should be the immediate short term aim of the atmospheric community.

We are still in the position where our knowledge of the processes taking place in the atmosphere is far from complete. Progress required in each atmospheric discipline is in many ways limited by progress in each of the other fields. Each field is ultimately a constituent part of a single system and hence co-operation is required. Some of the most critical issues relevant to the bureau of IAMAS, regarding the enhancement of our knowledge that need to be addressed over the coming decade are outlined below.

7.3.1 Atmospheric Chemistry

Underlying processes in the atmosphere are the chemical reactions taking place. The ability to measure and parameterise these microphysical processes is central to atmospheric science.

Clouds

Clouds are important in the atmosphere because they strongly modulate incoming solar and outgoing thermal radiation.

They are also the source of precipitation, and a key element in the hydrologic cycle. They are currently under intense scrutiny by researchers to gain a better understanding of their role in our environment *Liou* [1986]. This knowledge is needed for better predictions of both climate change, and forecasting.

Radiative properties of clouds depend upon their microphysical properties such as water-ice content, size spectra and shape. These same properties determine the precipitation rate *Dowling and Radke* [1990]. Understanding the role of these properties in geophysical processes and correctly parameterising the processes in models are essential. Cloud microphysics is a growing area of research, but currently limiting progress in other areas of atmospheric science (section 7.3.2). Forecasting and climate models both require improved parameterisations of clouds as the effect of spatial and temporal resolution on all processes and couplings in the atmosphere are investigated.

A number of investigations are under-way into different cloud types, including polar stratospheric clouds, notilucent clouds and cirrus clouds.

The ability to remotely sense cloud microphysical properties from satellites is also crucial if the cloud data is to be input into real time forecasting models with the global coverage that is required. Much research is occurring in the retrieval of cloud microphysical properties from remotely sensed data, e.g., *Baran et al.* [2001]. In addition, several missions, such as Cloudsat and ICESat, are in progress to measure cloud properties from remotely sensed data. With a combination of experimental and modeling initiatives, real progress will be made in what is undoubtedly one of the most complicated problems in atmospheric science.

Aerosols

Aerosol particles are also important in the atmosphere, directly by absorbing, emitting and scattering light and indirectly by serving as cloud condensation nuclei and impacting precipitation rates. They are considered a much greater uncertainty in climate forcing than CO_2 concentrations as they have a much greater spatial and temporal variability making their quantification more difficult. Their forcings are regionally variable and their couplings to climatic effects require significantly more research *Penner and et al.* [2001].

It is vital that aerosol distribution databases are both spatially and temporally resolved along with their sources and sinks. Further laboratory based study into the conversion mechanisms that transform precursors into aerosols, combined with 3-D global observations are required.

Existing systems, such as AVHRR, TOMS, GOES and METEOSAT, have been used to aid in improving the chemical meteorology need to aid aerosol estimates. New aerosol data products are also being obtained from SeaWiFS, MODIS, MISR and CERES instruments *NASA* [1999].

Ozone and Greenhouse Gases

Atmospheric abundances of most greenhouse gases are at their highest values since measurements began. Much publicised research over the past few years has demonstrated that ozone changes since the pre-industrial period are coupled to climate change. It is now the third most important greenhouse gas behind CO_2 and CH_4 *Ehhalt et al.* [2001]. The magnitude of its effect, however, is critically dependent upon the altitudes at which depletion occurs and must be thoroughly understood in order to unravel the detection of climate change from natural variability. It is understood that changes in all greenhouse gas abundances and lifetimes will have the effect of altering climate forcings, but uncertainties remain high.

These uncertainties are improving, due largely to collaboration in various observation and modeling campaigns. Better simulation of the past two decades, and of the current decade, will improve confidence in projections of greenhouse gases. Global observations from satellites, radiosondes, and surface data all remain essential in monitoring and understanding the chemistry, transport and feedbacks of ozone and other greenhouse gases. Currently, one of the most important problems that needs addressing is the modeling of sub grid scale processes in chemical transport models. The chemistry in the atmosphere is generally highly non-linear, and thus highly dependent upon resolution. Surface emissions are often highly localised so model resolution becomes critical. Improving the spatial scale in chemical transport models is hence essential. Techniques, which allow an increase in resolution locally via a two-way nesting approach, promise to provide the bridge between features from local to global scales.

The Carbon Cycle

Atmospheric chemistry is of substantial import in the understanding of the carbon cycle in terms of its reactions both in the atmosphere and at atmospheric boundaries *Prentice and et al.* [2001]. Carbon plays many roles in the Earth system. It forms our food and acts as our primary energy source but is also a major contributor to the planetary greenhouse effect and the potential for climate change (section 7.3.2). Carbon dioxide and methane concentrations have been increasing in the atmosphere, predominantly due to our use of fossil fuels and land clearing. Of the carbon dioxide emitted into the atmosphere, about half is currently taken up as part of the natural cycling of carbon into the ocean, and into land plants and soils. Successful carbon management strategies need to be based on solid scientific information of the basic processes affecting the global carbon cycle.

Models that include ecosystem distribution and condition, biological processes determining exchanges with the atmosphere from the land and the ocean, and the transport of carbon within the Earth system are required. These models will rely upon the integration of geographic information, remote sensing data and comprehensive databases.

The Global Water Cycle

Water plays integral roles in both surface conditions and the atmospheric circulation. The conversion of liquid and solid water to water vapour results in a local latent cooling. Water molecules also have a large impact on Earth's radiation budget. They are strong absorbers of infrared radiation, and the resulting greenhouse effect of atmospheric water vapour is, by far, the strongest determinant of the Earth's surface climate. Measuring and forecasting spatial and temporal patterns in water vapour and clouds are hence essential to addressing climate, water resources, and ecosystem problems. Aside from the previously mentioned problem of modeling cloud processes, two further areas of the hydrologic cycle that require improvement are in the subgrid scale modeling of soil moisture and evaporation and the coupling of these models to GCMs.

The global water cycle interacts with all components of the Earth system. A hierarchy of simulation models are required, including GCMs, mesoscale models, basin scale hydrologic models along with validation against global observation data such as EOS. Techniques utilising the recent advances in computing power will allow for rapid progression in the modeling of the hydrological cycle.

Summary

The parameterisation and modeling of cloud microphysics and the high resolution modeling of chemical transports are currently both vital areas of research . With recent advances in computing power, both climate and forecasting models are experimenting with optimal methods for faithfully representing cloud processes and chemical transports. The research objective of attaining sufficiently accurate parameterisations can only be achieved through a synergism of transport models and GCMs validated against comprehensive spatially resolved databases. The implementation of global observing systems measuring horizontally and vertically resolved critical cloud and chemical parameters will be required as input to both the models and the databases. These measurements should involve a combination of remote sensing, in-situ, aircraft and radio sonde data. Research will certainly be helped by observations on improved spatial scales, such as HIRDLS (*Gille et al.* [1994]). In conjunction, a systematic effort is required to improve the documentation of time resolved atmospheric composition over regional and global scales. This is a book keeping exercise but imperative for the monitoring of pollution and climate change.

7.3.2 Climate and Climate Change

The aim of understanding the physical and chemical basis of climate and climate change in order to predict climate variability from timescales of seasons to centuries and beyond and to assess the role of human activities in affecting climate is essential to our continued well-being. Directly coupled to climate are factors central to sustaining life on the planet, such as water resources, food and temperature. For a fuller discussion on climate and climate change see *IPCC* [2001].

Development and Integration of High Resolution Coupled Climate Models

Numerical models of the Earth system must be based upon an accurate representation of physical, chemical and biological processes. All processes and interactions, which occur on a wide range of spatial and temporal scales and, in many cases, are highly non-linear, need to be included.

Modeling the climate system will require two basic strands of development: improvements in model resolution and improved model parameterisations. The first of these allows us to more realistically model finer scale phenomena, with the second allowing us to better deal with those scales we cannot model explicitly.

It has long been known that small scale features can manifest an influence upon large scale processes in the atmosphere on finite timescales. This has probably been most memorably stated by Lorenz, "...one flap of a seagull's wing may forever change the course of the weather." It is imperative to develop models that can explicitly simulate flows on smaller scales, enabling the capture of significant non-linear interactions between different space and time scales and different components of the Earth system. Whilst there will always be unresolved scales of motion, giving rise to a finite forecast horizon, at the present time, an increase in resolution would greatly improve the value of predictions.

Most current climate models have a typical resolution of $\sim 3°$ in the atmosphere and $\sim 1°$ in the ocean. This does not allow for adequate representation of the climate system in either component. An eddy permitting ocean model has been shown to provide a better representation of western boundary currents and a more accurate simulation of equatorial waves (which are a key element of El Nino). Higher resolution simulations of the atmosphere have also already demonstrated improvements in regions where orographic effects are important *Pope and Stratton* [2002]. Both chemical and land surface effects are again highly dependent upon resolution and would benefit from resolving finer scales (sections 7.3.1 and 7.3.2). Higher resolutions of $< \sim 1/3°$ ocean and $< \sim 1°$ atmosphere will allow the exploration of the processes in climate system that give rise to its variability on all scales.

Even with increases in resolution, climate modellers still also face the on-going need to improve parameterisations of sub-grid scale phenomena. Effects from arbitrarily small scales will still affect behavior at larger scales *Lorenz* [1969], and the climate modeling community needs to address better ways of dealing with the residual incompleteness of models. Cloud microphysics, for example, will need to be parameterised regardless of any foreseeable improvements in resolution.

Within any general circulation model (GCM) there are many parameters which are poorly constrained by observations. Many of these have been found to be sensitive to small perturbations. Parameter perturbations can add non-linearly, and in order to better understand the physics of the models we need to disentangle the physical mechanisms behind model response. We need to identify those processes which have little impact on the mean climate but a large effect on climate sensitivity, and secondly identify those parameterisations which are both influential and poorly constrained. The first of these undertakings will help us gain an improved understanding of climate sensitivity and second will address systematic model errors. As we develop more convincing and higher resolution component models, our reliance on parameterisations will broaden and there will be a continual need to validate these parameterisations.

Recent advanced parallel super computers (such as the Earth Simulator in Japan) will enable the development and utilization of very high resolution climate models and very complex earth system models including atmosphere, ocean, land surface, chemistry, biosphere, cryosphere and so on. Such models will undoubtedly open the door to a new area of study in the earth sciences and lead to new findings and understanding as results from the next generation of instruments measuring the atmosphere are established.

Improve Understanding of Anthropogenic Effects upon the Atmosphere

Global climate change has attained enormous interest due to its potential to effect human activities and the environment. A critical issue is to understand the possible anthropogenic role in climate change.

Human-induced changes in climate observed on decadal time scales are already evident. Atmospheric pollution now hangs over many regions of the Northern Hemisphere. The thinning of the stratospheric ozone layer, most notably over Antarctica, is a human-caused feature of the planet. The greenhouse gases being added to the atmosphere will reside there for decades to centuries and are predicted to increase average global surface temperatures by several degrees, a change that is larger than the natural variation occurring over the past 15,000 years. Precise understanding of the effects of human influence on the environment and the communication of theses results to policy makers is essential. The separation of natural climate change from that caused by the activities of humans needs to be established before the onset of a dynamic that is irrevocable.

Research is continuing into the coupling between climate and anthropogenic influences, e.g., *Giorgi and et al.* [2001]. Our anthropogenic emissions are often from highly localised sources and experience highly non-linear interactions. We are subjecting our planet to forcings to which it has never be-

fore been exposed and the repercussions need definitive evaluation. Only through the development and application of a high resolution integrated global modeling and observation system can our influence on the climate and environment be estimated. A better understanding of sources, processes and couplings between different parts of the climate system is required so that anthropogenic effects can be accurately and efficiently represented in models. Data from the next generation of Earth Observing Satellite programs (section 7.4) and the use of ultra high resolution computer models will undoubtedly aid in the unraveling of our effects upon the environment.

Atmospheric Science in Environmental Issues

The atmospheric sciences are a vital component in the control of environmental issues, e.g., in the monitoring of pollution and greenhouse gases. As certainty in causes of climate change improves, the thresholds of tolerable risks due to projected climate change need to be defined and the communication of scientific results to both society and policy makers improved. The management of water resources is another environmental issue in which results from atmospheric physics research have an integral role to play.

In both cases our understanding of the processes involved needs furthering so that we are in a position to convince policy makers. Policy makers have many factors to consider in their decisions and policy changes cannot be expected until we can provide them with results based upon reliable prediction.

Regionalisation of Climate Change Projections

As the climate changes, there will be significant differences between regions *Giorgi and Mearns* [1991]. Climate predictions, hence, require regionalisation, in terms of impacts and vulnerability.

Regional climate is often affected by forcings and dynamics on a sub grid coupled Atmosphere-Ocean GCM (AOGCM) scale. Much of the current difficulty lies in extracting fine scale regional information from coarse scale AOGCMs. Improvements in the resolution of climate models as computing power increases will also aid in the regionalisation of climate change predictions.

Land Surface Effects on Climate Variability

Understanding land-atmosphere interactions and their effect on climate variability is an important area in climate modeling. The effect of the coupled terrestrial biosphere and global atmosphere on climate change requires investigation, particularly with regard to its effect on the hydrologic cycle, heat and momentum fluxes and increases in atmospheric CO_2 *Entekhabi et al.* [1992]; *Randerson et al.* [1997]. Variability in properties such as topographic height, vegetation cover, soil properties and snow cover occur at all length scales and combine non-linearly. The accurate modeling of surface processes is thus strongly constrained by horizontal resolution. Explicitly resolving finer scale features will lead to many improvements in the simulation of climate. The use of remote sensing data from missions such as Terra will also significantly enhance current knowledge in concert with data from other land surface experiments.

Climate and the Ocean

The ocean plays a critical role in the physical and biogeochemical dynamics of the ocean/atmosphere system.

Through heat exchange and freshwater transports, the oceans perform an active role in driving longer term atmospheric variations. The ocean also has a significant role in biogeochemical dynamics, particularly in the carbon cycle due to the dissolution of CO_2 in water and the release of dimethylsulphide by plankton. Climate is additionally hugely effected by sea-ice due to its consequences for ocean water masses. Correct parameterisation of this is needed, which requires high resolution atmospheric and oceanic fields.

The ocean feedback to climate must ultimately be manifested through air-sea exchanges and these interactions require further study. Some phenomena over the ocean that require study are the vertical exchanges of heat, moisture and energy across the planetary boundary layer, boundary layer clouds, tropical convective clouds and midlatitude convective systems.

Coupled climate models certainly require an increase in resolution from $\sim 1°$ to $<\sim 1/3°$. This resolution increase is required in order to adequately capture eddies in the ocean. Planetary Rossby waves are also a key component of ocean circulation and may play a critical role in setting the timescale for coupled modes of variability. Rossby wave propagation speed and amplitude could be very different in ocean models with different resolutions. Models hence need to be tested in comparison to available data.

Better historical heat content estimates for the ocean would improve Ocean General Circulation Model (OGCM) validation. New remote sensing data and in-situ measurements (for probing deeper into the ocean) will also augment OGCM performance and thus the understanding of the of the oceans in climate. Instruments, such as TOPEX/Poseidon, are used to study large scale circulations, infrared sensors, such as the Advanced Along-Track Scanning Radiometer (AATSR) provide data on sea surface temperature, Scatterometers, such as Seawinds, supply sea wind data, passive microwave radiometers, such as AMSR-E, are used to calculate latent heat fluxes and ocean color sensors such as SeaWiFS, provide basic information on biological processes. These data sets, when combined with models, will allow a systematic investigation of the coupled atmosphere-ocean system.

The ocean is dealt with in considerably more detail by the bureau of IAPSO.

Improve Uncertainties in the Hydrologic Cycle

Deficiencies in our understanding of the global water cycle severely handicap efforts to improve climate prediction and guide water resource planning. Significant problems in climate modeling are underpinned by our limited ability to understand and model cloud processes (section 7.3.1) and the hydrologic cycle *Stephens et al.* [1990]. Clouds are strong absorbers of infrared radiation, thus driving the greenhouse effect. Water vapour is the largest component of the greenhouse effect, and hence the strongest determinant of the Earth's surface climate. Changes in precipitation distributions are likely to have significant impacts on future climates, especially in the form of water stress. Any adverse effects can be expected to be borne by those countries already facing serious water resource constraints. Model prediction of changes to the global hydrologic cycle are still not reliable due to the non-linear response of precipitation to changes in forcing/temperature. This is illustrated in figure 7.3.2. It is clear that trends in global-mean temperature are predicted much more accurately than trends in precipitation. It is still proving difficult to isolate the anthropogenic contribution to changes in the hydrologic cycle, which seems to be dominated largely by natural forcing. Measuring and forecasting spatial and temporal patterns in water vapour and clouds are hence essential to address climate, water resources, and ecosystem problems *Allen and Ingram* [2002]; *Lindzen* [1994].

Research is in progress in trying to improve our knowledge of cloud processes and the water cycle (section 7.3.1). Through these undertakings, the urgent requirement for uncertainty reduction in climate prediction will be greatly aided.

The Carbon Cycle and Climate

Carbon cycle feedback has a very important influence on the climate. The magnitude of this feedback is uncertain and depends strongly on the response of soil respiration to temperature, ocean exchanges and other processes, all of which require research.

The Effects of Solar Variability on the Global Climate System

The effects of solar variability on the climate may be significant and must be differentiated from other natural influences and from climate effects associated with human activity.

Improved understanding of solar influences on the tropospheric and stratospheric circulation and the climate in general is needed *Labitzke and et al.* [2002]. Several experiments such as Earthshine (proposed) will greatly help in characterising the solar influence on our atmosphere and climate *Lockwood* [2002].

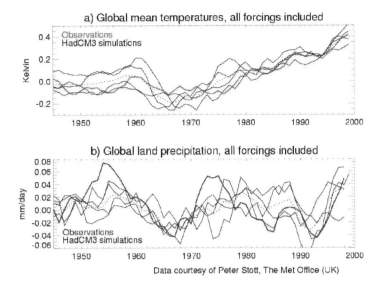

Figure 7.1: Changes in observed global-mean temperature and land precipitation over the last 55 years compared with a climate model. After *Allen and Ingram* [2002].

E-science Enterprises

Current and anticipated advances in network technologies are opening up new avenues for scientists. Chief among these is the development of e-science or GRID technologies. GRID projects are based on making use of high speed network connections, distributed data and software to enhance scientific research.

Two very different e-science projects, which perhaps give an example of the breadth of these emerging technologies, are the Earth System Grid [see http://www.earthsystemgrid.org/] and climateprediction.net [see http://www.climateprediction.net/]. The Earth System Grid seeks to build a community wide resource from very large sets of climate model data and analysis software, enabling climate scientists to perform research more efficiently and at a greater depth. Climateprediction.net, on the other hand, makes use of distributed computing techniques to produce a very large ensemble climate forecast, generated by perturbing both model parameters and initial conditions. The general public will shortly be invited to participate in climateprediction.net, performing pre-packaged (10-100 year) simulations with the ultimate aim of establishing the spread of forecasts of 2050 climate consistent with current observations and recent climate change.

These are just two among many exciting projects, which seek to exploit emerging technologies to enable us to enrich our scientific knowledge in new ways.

Summary

The overarching short-term aim of climate modeling is to reduce the uncertainties (beyond those inherent in the system) in climate prediction, hence increasing the value of the predictions to society. Uncertainties must be quantified, allowing useful communication of information to end users and policy makers.

Construction and evaluation of higher resolution models that are increasingly comprehensive, incorporating all the major components of the climate, are needed as we aim to predict the future climate with greater accuracy. Current resolutions do not represent many processes within the climate system sufficiently well. In addition to improvements in resolution, there needs to be a complimentary set of developments in the parameterisations included in models. Recent advances in computing power should facilitate the construction of higher resolution models with the prime focus for research being the continuation and expansion of efforts to elucidate key climate processes, both chemical and dynamical.

The creation of a permanent global climate observing system is required in order to improve climate change prediction and thereby allow for regionalisation. This can only be achieved if the major global climate forcings and radiative feedbacks are monitored with adequate precision. This would allow for tighter constraints on possible trends in future climate. The extension of the observational climate record through the development of integrated historical and current data sets will also contribute by allowing a greater probabilistic certainty in climate prediction. This could also be aided by comparative planetology.

There is still great uncertainty in climate prediction but

significant progress is being made as the underlying physics of the atmosphere is better understood and implemented into models *IPCC* [2001].

7.3.3 Atmospheric Dynamics and Meteorology

Along with chemical processes in the atmosphere, dynamics provides the building blocks of the atmospheric sciences. Improvements in our knowledge over the last decade have lead to dramatic reduction in forecast errors over all timescales. This field of research is key to our quality of life and everyday comfort. The capabilities of meteorology are already impressive but there exist many areas that allow for improvement.

Forecasting

Reliable forecasting is becoming increasingly important to society especially as regards its effect upon energy, health and agriculture. Accurate weather forecasting allows a greater certainty when planning ahead. It also benefits the management of water resources and other elements having a direct influence upon the sustainability of both our environment and our economies.

The skill in forecasting has improved dramatically over the last 20 Years, and today, we enjoy considerable confidence in predictions from periods of days to years. Other, more advanced, measures of forecast skill exist today, but as an illustrative example, figure 7.2 shows the increase in forecast skill for 500hPa height forecasts. Whilst forecasting skill has improved dramatically, the useful range of forecasts is currently of the order of a week. Extension of this range from weeks to months is the next hurdle in forecasting. In order to achieve this, the resolution of models will need to be improved in order to give a better representation of physical processes, particularly in the coupling of the ocean, atmosphere and land-surface. Progress in numerical weather forecasting will be made as improvements in the way that physical processes are treated are improved and then incorporated into the models. Some models are now fully non-hydro static and it is the parameterisation of sub grid scale processes that require most research. New models for rainfall accumulation, surface hydrology, convective weather and cloud diagnosis have all aided in improving forecasting performance *MetOffice* [2002]. Prediction and modeling of cloud cover and cloud condensate certainly needs improving. New schemes for the temporal evolution of cloud fraction and condensate are being developed but require more research (section 7.3.1).

Many problems in forecasting will also benefit from proposed developments in the processing and assimilation of satellite data, especially in the direct assimilation of radiances into models. Meteorological observing systems continue to expand and improve as observations throughout the world are becoming more thorough and co-ordinated through undertakings such as the Working Group for Planning and Implemen-

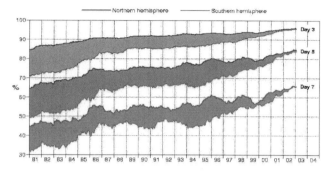

Figure 7.2: The evolution of forecast skill for 500hPa height forecasts, for 3, 5 and 7 day forecasts for both the Northern and Southern Hemispheres. Figure reproduced with kind permission from ECMWF.

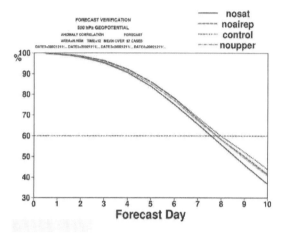

Figure 7.3: The relative importance of different observation techniques in forecasting. The effects of the removal of satellite (NOSAT), aircraft measurement (NOAIREP) and radiosonde (NOUPPER) data sets from a forecast of 500hPa height are shown. Figure reproduced with kind permission from ECMWF.

tation of World Weather Watch. Many surface observations are becoming automated with significant success. Observation times are also being reduced. Remotely sensed data is now by far the most significant data set in forecasting. Figure 7.3 shows the effect on forecasts caused by the removal of certain data sets.

The importance of remotely sensed satellite data is clear. New satellite programs such as Meteosat Second Generation (MSG) and EUMETSAT Polar System (EPS) will further help in improving forecasts. Additionally, Radio-sonde and aircraft data remain highly important, especially for wind and relative humidity measurements, and must not be allowed to deteriorate.

Synoptic Meteorology

Recent research has dealt with such diverse subjects as large-scale tropical and subtropical disturbances, extratropical cyclones, polar lows, the interactions between tropical and extratropical systems, and the large-scale effects of volcanic eruptions. The understanding and modeling of these large scale effects will improve in direct correlation with the resolution and integrity of atmospheric models.

Improve Predictability of Extreme Weather Events

Extreme weather events are at the heart of some of the planet's most devastating natural disasters. The ability to predict these events and minimize their effects is vital. Progress in the study of atmospheric interactions that shape weather phenomena has created the opportunity to make major advances that have lead directly to improved weather warnings and predictions. This gives communities a greater confidence, an improved chance of minimizing the effects of any extreme weather events and hence a superior quality of life. New models have the capability to run at horizontal resolutions of around 1km or less and open up the possibility of significantly improving predictions of severe weather. Increases in computing power will probably be required before these high resolutions become fully operational. Elucidating relationships between extreme events and modes of climate variability, such as ENSO, will also be vital.

Atmospheric Dynamics Related to Climate Change

Several processes and feedbacks within the atmosphere require improved understanding. ENSO is the largest of the natural climate fluctuations and is now qualitatively understood, but quantitative understanding is still weak, mainly because it involves interaction between sea surface temperature anomalies and organised convection. It requires continued monitoring due to its potential social and economic implications. Other important processes are the North Atlantic Oscillation (*Hurrell* [1995]), the Arctic Oscillation and the Madden Julian Oscillation. There are many possible mechanisms at work and continued observations and modeling efforts are required to illicit the actual feedbacks responsible.

Research is also required into the dynamics and transports in the tropics. The tropics act as heat source regions that drive global atmospheric circulation. Understanding short-term changes in our climate requires an improved understanding of the mass and heat balances of the tropics. For example, what processes maintain the western Pacific warm pool, and what is the role of El Nino Southern Oscillation (ENSO) events in this maintenance? What role is played by surface heat fluxes, and what role is played by tropical to extra-tropical exchanges?

Research in the area of climate dynamics will be helped by new observations and by high resolution simulations.

Global Changes in the Middle and Upper Atmosphere

Global changes in the middle and upper atmosphere, in response to both natural and anthropogenic influences, have significant effects on the lower atmosphere. The middle atmosphere is a highly coupled system in which dynamical, radiative and chemical processes are intimately connected. Important radiative roles are played by CO_2 and O_3. Exchanges with the troposphere and higher atmospheric levels are also important, and it is vital that these processes are understood. An approach involving data from satellite programmes, such as EOS, and sophisticated models are required to understand circulation, chemical, dynamical and radiative processes and horizontal and vertical coupling mechanisms.

Boundary Layer Dynamics

Most dynamical processes driving global change relate to physical-chemical-biological linkages, and exchanges of energy and materials across three main interfaces: between land and ocean, ocean and air, and air and land. As models are advanced and the effect of both horizontal and vertical resolution upon processes is established, it could well transpire that greater vertical resolution is required in surface layers and the atmospheric boundary layer in order to represent processes accurately. In order to improve these models, the dynamics and transports associated with exchanges in the atmospheric boundary layer will need further investigation. Increases in the vertical resolution of observations at interfaces might be needed to validate models.

Convection

Convection plays an important role in the surface heat budget and whilst understood, is poorly characterised in most models. Deep convection and stratosphere troposphere exchanges have a great effect on the atmospheric vertical profiles of trace gases and aerosols. Organised convective transport processes develop on small scales but are globally important. These smaller-scale hydrodynamic effects and non-hydrostatic motions are typically poorly represented in most GCMs. CO is an excellent tracer of convective transport and CO data from several satellites will help in improving current convective models.

Atmospheric Electricity

Although the existence of the global electric circuit has been known for many decades, most recent research has concentrated on the 'generators,' i.e., the thunderstorms. There are many knowledge gaps concerning, for example, the contribution of other generators, such as the solar-wind and the upper atmosphere tides.

Projects, such as the U.K. Met Office long-range lightning detection system, will allow the original position of lightning discharge to be located with an accuracy of between 2 and 5km over most areas of Europe. In addition, experiments studying the sun, such as SOHO, and those studying the interactions of the sun with the thermosphere, such as TIMED [see http://www.timed.jhuapl.edu/], will allow for improved characterisation of the causes and effects of atmospheric electricity.

Understanding the Nature and Characteristics of Turbulence in Rotating Fluid Motion

Underpinning the study of atmospheric dynamics is the study of fluid flow and particularly turbulent motion. Further, laboratory-based study of the dynamical processes occurring in the atmosphere is needed alongside the continued modeling efforts.

E-science and Computer-to-Computer Communication

Weather-dependent enterprises could incorporate atmospheric information more readily into their decisions with greater computer-to-computer communication and increased ease of access to electronic resources. The atmospheric sciences should aid in this process wherever possible. The world data center system, created through the international council of scientific unions, is one such undertaking that benefits atmospheric scientists and could be used as a model for further initiatives. The meteorological community also relys upon fast exchange of observational data on a national and international scale. Use of internet protocol standards has already allowed for much speedier and reliable exchanges of data. It is important that the atmospheric sciences remain up to date with advances in communications and utilise all of these in the progression of research.

Summary

Better utilization of data, including the determination of optimal combinations of available and new observations are needed to enhance research in atmospheric dynamics and meteorology. The deterioration of the global radio sonde network and other in-situ measurements needs halting. The advances in satellite programs will also continue to assist in further research in dynamics and meteorology. New satellite instruments will provide data with global coverage and a vertical resolution that will be essential as modeling exercises move to higher resolution in order to model the processes in the atmosphere. Furthermore, observations and understanding of boundary atmosphere interactions will be key in improved forecasts of convection, precipitation, and seasonal climate.

All of these efforts will likely be benefitted by e-science enterprises and the improvements in computing capacity in the next few years. As computing capacity increases, so models, parameterisation and observations must be ready to fully exploit the advances.

7.3.4 Planetary Atmospheres

Other planets in the solar system and beyond provide atmospheric scientists with full scale laboratories in which to test dynamical, chemical and climate theories. Missions proposed over the next few years will provide more data to improve planetary GCMs and aid in comparative planetology.

Establish Climate Histories and GCMs of Earth, Mars, Venus, and other Planets

Regarding the aim of a sustainable Earth, the atmospheric sciences can also contribute to understanding how our planet is likely to develop over the coming centuries through comparative planetology. This provides an excellent base for the study of other atmospheric regimes and provides additional understanding on how our planet might develop. In the reconciling of GCMs of the terrestrial planets, significant information can be attained, particularly in terms of greenhouse (radiative energy balance) models, that can help determine the stability of past, present and future climate on Earth. The gas giant planets also pose noteworthy research problems for the future.

In order to facilitate the understanding of other planetary systems, a greater emphasis must be placed upon planetary missions. Systematic measurements are required, utilizing both orbiters and landers and will pose problems from both a scientific and engineering perspective. Missions proposed over the next decade will begin to address this lack of data for both Mars and Venus.

Climate Conditions for Life Elsewhere in the Galaxy

Assessing the possibilities for life elsewhere in the Galaxy is a growing area of interest and is important if the goals of section 7.2 are to be realised. Several projects looking for extra solar planets are currently in progress.

Summary

An increased number of planetary missions and greater emphasis on comparative planetology will afford much additional insight regarding the Earth and provide the foundations for development of new habitable environments on other planets .

7.4 Recommendations for the Atmospheric Sciences

Recommendations for Moving Towards Higher Fidelity Models

Progress toward the goal outlined in section 7.2 is accelerating rapidly as the atmospheric science community is beginning to develop techniques to utilise the vast increases in computing power that are becoming available today. The realisation that as greater computing power becomes available, the best way to exploit this is not just to run current models for longer, or more often, but to improve the fidelity, spatial and temporal resolution of models, is going to lead to huge developments in the atmospheric sciences in the next few years.

The linkage between the Earth and its atmosphere is the ultimate symbiotic relationship. Atmospheric studies are being shaped today by the recognition that contemporary approaches must seek to understand, model, and predict all components of the Earth's environment, including coupling between various systems. Critical scientific questions focus on the exchanges of energy, momentum, and chemical constituents between the troposphere and the surface below, as well as with the layers above. In addition, many processes within the atmosphere itself, such as cloud processes and chemical transport, are not adequately represented in current models. Parameterising all of these processes and elucidating their resolution dependence, along with an understanding of processes coupling different spatial and temporal scales, will be the primary research focus over the next decade.

The over-riding recommendation for the atmospheric sciences is hence to improve co-ordination and co-operation in the development of high resolution Earth System Models both for climate research and short-term forecasting.

A few other general recommendations for the atmospheric sciences, aimed at moving toward the development of higher fidelity models and ultimately the realisation of the goal outlined in section 7.2 are given below.

- Improve Communication and Co-ordination with Scientists Studying other Elements of the Earth System

  As models progress to greater resolutions, so better simulation of the complex behavior of the full Earth system is required. It is the fluxes of materials and energy between the atmosphere and the oceans, the land, the ecosystems, and the near-space environment that define the structure and evolution of atmospheric processes and events. Most goals of research in the atmospheric sciences are likely to require considerable collaboration with a broad range of other geoscientific disciplines. By moving toward higher resolution in atmospheric models, the gap will be narrowed between climate modellers and scientists specialising in other aspects of the Earth system, so aiding better integration.

In developing Earth system models, subsystem models on vastly differing scales, both temporal and spatial, must be reconciled and combined. In the testing and integration of these models, much greater co-ordination and co-operation is required between fields within the Earth sciences. SOLAS (Surface Ocean-Lower Atmosphere Study) is an example of an interdisciplinary and international investigation of the dynamic processes occurring across the sea surface with the IGBP (International Geosphere Biosphere Programme) co-ordinating several other efforts in the research of boundary-atmosphere interactions. These types of undertakings will be essential to accelerated progression.

- Use of Contemporary Numerical Computer Models

  The atmospheric sciences must ensure that the most contemporary computing facilities are used along with models that fully utilise the capacity available. The parameterisations required for higher resolution investigations should be developed in advance of increases in computing power so that they can be fully exploited as they arise.

- Integrated Global Observing Systems

  The increasing importance of integrated observation systems and new observations of critical variables is clear. Ultra-high resolution satellite data is becoming important as models progress toward greater resolution. The global coverage possible from satellites is vital, but high resolution in-situ measurements are also needed, particularly for interactions and exchanges at the boundaries of the atmosphere. An increase in the number of surface exchange measurements will aid research, especially if the vertical resolution of models at boundaries is enhanced. The atmospheric science community should develop a plan for optimizing global observations of the atmosphere, oceans, and land. This plan should take into account requirements for monitoring weather, climate, and air quality and for providing the information needed to improve predictive numerical models used for weather, climate, atmospheric chemistry, air quality, and near-Earth space physics activities. Both NASA's Earth Science Enterprise and ESA's Living Planet program are significant steps toward reaching this goal NASA [1999]; ESA [1998]. Development of commercial remote sensing ventures might also be considered.

- Data Assimilation

  The maximum amount of information must be extracted from observations. This requires continual adaptations and advancement of assimilation techniques. The quality of both observations and modeled data needs thorough assessment in optimal assimilation techniques. Current systems such as the 4-D Var system of the U.K.

Met. Office provide an excellent model for assimilation into forecasts. Efficient assimilation of data into climate models requires similar thought.

- Laboratory Based Study

  In elucidating key chemical processes and transports in the atmosphere, laboratory based study of chemical and dynamical processes needs to be continued and enhanced. As models begin to require improved parameterisations on higher resolution grids, so tests of these parameterisations in the laboratory will be needed as an integral part of the validation process.

- Database Management

  One of the limiting factors associated with climate models at the present time is data with which to validate results. A concerted effort in the development and maintenance of spatially and temporally resolved chemical and climatological databases will be central to the validation of models in the future.

- Mathematical Techniques

  Perhaps the most challenging aspects of developing higher resolution models are the mathematical techniques required to bridge the gaps between disparate spatial and temporal scales associated with different processes in the atmosphere and at its boundaries. Adaptive gridding schemes, which employ a combination of nested static and dynamic grids, which can each run at their own optimal timestep, promise to significantly increase the performance of models. The development and validation of mathematical techniques is vital.

- Increased Emphasis on Comparative Planetology

  In understanding processes in the atmosphere and modeling climate changes, there is a vast amount that could be gained from an increased use of comparative planetology and exchange of results between planetary and Earth scientists. As models of processes within the Earth system are improved to higher resolution, the ability to compare the effects of different parameters on couplings within the atmosphere that is provided by comparative planetology will be invaluable.

- Greater Use of E-Science

  In the modern era, a greater use of the freedom of information provided by the electronic age would be invaluable in the atmospheric sciences.

Recommendations for Realising Long-Term Projects

Whilst we are not currently in the position to realise the long term projects outlined in section 7.2, significant progress can already be made, and the foundations for future research established. Through the undertaking of ambitious projects like these, real progress is likely to be made in the basic understanding of atmospheric processes. In challenging ourselves and taking a truly long-term view, not only will great strides be made toward achieving the conclusion of the projects themselves, but new perspective and insight into more short term problems will also be revealed. Conservatism has never been a friend of research.

- Intentional Forcing of the Atmosphere

  It would be irresponsible in the extreme to begin forcing our atmosphere at the present time with our limited knowledge of exactly what the consequences might be. However, this is something that society should certainly consider in the future. Being able to modify the chances of precipitation or null the onset of extreme weather would certainly be of benefit to society. Research should be starting now with both computer and laboratory based simulations. The study of point-like forcings, such as volcanic eruptions and anthropogenic emissions, will also aid in the development of our ability to effect a positive influence of aspects of our environment.

- Location and Development of Alternative Habitable Environments

  As previously mentioned, several projects are already underway into the detection of Earth-like planets around other stars. This type of undertaking is vital along with increased investment in the engineering capabilities required to inhabit another planet. In addition, establishing terraforming projects would possibly be of huge benefit to both society and the atmospheric community. Through these types of projects, significant information will be returned about different couplings within the Earth system along with detail about how to maintain a habitable environment, which has to be our ultimate aim.

Other Recommendations for the Atmospheric Sciences

- Increase Public Information and Communication

  The atmospheric sciences should improve presentation and communication of results to society and policy makers. We are all subjected to much misinformation about the dangers of climate change *Lomberg* [2001]. This is particularly important when relating details of extreme weather events to the general public. Uncertainties on projections of weather and climate need to be quantified and properly represented to the public and policy makers.

7.5 Conclusions

Many important and challenging opportunities are facing the atmospheric sciences. As the resolution and fidelity of atmospheric models and observations improve and atmospheric processes are better understood, the accuracy and resolution of atmospheric prediction will be enhanced. As a consequence, society will enjoy greater confidence in atmospheric information and will be able to act more decisively and effectively. Through the adoption of a forward thinking approach the IUGG will have a significant role to play the realisation of this vision.

Bibliography

Allen, M., and W. J. Ingram, Climate and water, *Nature Insight*, *419*(6903), September 2002.

Baran, A. J., P. N. Francis, L. C. Labonnote, and M. Doutriaux-Boucher, A scattering phase function for ice-cloud: tests of applicability using aircraft and satellite multi-angle multi-wavelength radiance measurements of cirrus, *QJR Meterol Soc*, *127*(2), 2395–2416, 2001.

Dowling, D. R., and L. F. Radke, A summary of the physical properties of cirrus clouds, *J. Appl. Meterol.*, *29*, 970–978, 1990.

Ehhalt, D., M. Prather, and et al., Atmospheric chemistry and greenhouse gases, In Houghton, J. T., and et al., editors, *Climate Change 2001: The Scientific Basis*. Cambridge University Press, 2001.

Entekhabi, D., I. Rodriguez-Iturbe, and R. L. Bras, Variability in large-scale water balance with landsurface atmosphere interaction, *J. Climate*, *5*, 798–813, 1992.

ESA, , The science and research elements of ESA's living planet programme, Technical report, ESA, 1998.

Gille, J., J. Barnett, M. Coffey, W. Mankin, B. Johnson, M. Dials, J. Whitney, D. Woodward, P. Arter, and W. Rudolf, The High Resolution Dynamics Limb Sounder (HIRDLS) for the Earth Observing System, *SPIE*, *2266*, 330–339, 1994.

Giorgi, F., and et al., Emerging patterns of simulated regional climatic changes for the 21st century due to anthropogenic forcings., *Geophys Res Lett*, *28*(3), 3317–3320, 2001.

Giorgi, F., and L. O. Mearns, Approaches to the simulation of regional climate change:a review, *Rev. Geophysics*, *29*, 191–216, 1991.

Hurrell, J. W., Decadal trends in north atlantic oscillation regional temperatures and precipitation, *Science*, *269*, 676–679, 1995.

IPCC, , Climate change 2001: The scientific basis, Technical report, IPCC, 2001.

Labitzke, K., and et al., The global signal of the 11-year solar cycle in the stratosphere: observations and models, *J Atmos Sol-Terr Phys*, *64*, 203–210, 2002.

Lindzen, R. S., Climate dynamics and global change, *Annu. Rev. Fluid Mech*, *26*, 353–378, 1994.

Liou, K. N., Influence of cirrus clouds on weather and climate processes: a global perspective, *Mon. Weather Rev.*, *114*, 1167–1199, 1986.

Lockwood, M., The earthshine mission, Technical report, Rutherford Appleton Laboratory, 2002.

Lomberg, B., *The Skeptical Environmentalist*, Cambridge University Press, 2001.

Lorenz, E. N., The predictability of a flow which possesses many scales of motion, *Tellus*, *21*, 298–307, 1969.

MetOffice, , Scientific and technical review, Technical report, U.K. Met Office, 2002.

NASA, , EOS reference handbook, Technical report, NASA, 1999.

Penner, J. E., and et al., Aerosols, their direct and indirect effects, In Houghton, J. T., and et al., editors, *Climate Change 2001: The Scientific Basis*. Cambridge University Press, 2001.

Pope, V. D., and R. A. Stratton, The processes governing horizontal resolution sensitivity in a climate model, *Clim. Dyn*, *19*, 211–236, 2002.

Prentice, I. C., and et al., The carbon cycle and atmospheric carbon dioxide, In Houghton, J. T., and et al., editors, *Climate Change 2001: The Scientific Basis*. Cambridge University Press, 2001.

Randerson, J. T., M. V. Thompson, I. Y. Fung, t. Conway, and C. B. Field, The contribution of terrestrial sources and sinks to trends in the seasonal cycle of atmospheric carbon dioxode., *Global Biogeochem. Cycles*, *11*, 535–560, 1997.

Stephens, G. L., S. C. Tsay, P. W. Stackhouse Jr., and P. J. Flatau, The relevance of the microphysical and radiative properties of cirrus clouds to climate and climate feedback, *J. Atmos. Sci*, *47*, 1742–1753, 1990.

Chapter 8

International Association of Geomagnetism and Aeronomy

A. J. Ridley

The International Association of Geomagnetism and Aeronomy (IAGA) is focused on understanding the near-Earth space environment. This ranges from studying the dynamics of the core magnetic field of the Earth, to the chemistry, dynamics, and electrodynamics of the thermosphere and ionosphere, to the interaction between the Earth's magnetic field with the Sun's atmosphere and magnetic field, to the Sun's atmosphere, and finally to the Solar dynamics which ultimately drive most of the science in the near-Earth space environment.

8.1 Long-term Goals

In order to lead an international effort on the understanding of this extremely large physical system, it is important to have a long-range strategy outlined. This long-ranged strategy helps to keep the research community focused on the central goals.

8.1.1 Focus Areas

Because IAGA includes such a diverse community, there are a number of long-term goals. Most of them can be described as having a complete understanding of the near-Earth space environment. Specifically,

1. Understanding the mechanism for the reversal of the Sun's magnetic field every 11 years and the flux emergence from the convective zone into the corona is of primary importance. While this process is far away from Earth, it sets up the solar wind structure and therefore plays a crucial role in determining the magnetospheric and ionospheric conditions. Understanding these processes would allow solar physics to begin to predict when and where coronal mass ejections may occur, where corotating interaction regions may be, and other

processes which have a dramatic impact on our space environment.

2. Understanding reconnection. While this process is somewhat understood in general terms, the specifics are not understood enough to predict when and where reconnection will take place in 3D. This means that a number of processes on the Sun and in the Earth's magnetosphere can not be fundamentally understood. For example, the explosive reconnection in the Earth's magnetotail (i.e., substorms), or the explosive initiation of coronal mass ejections on the sun are not well understood at this time. Once reconnection is understood thoroughly, substorm and CME initiations will most likely be understood.

3. Cross-scale processes in the ionosphere and thermosphere. It is known that auroral arcs can be very narrow and intense, but it is not known what effect those arcs have on the regional or global structure of the ionosphere. In addition, the entire high-latitude energy deposition is known to influence the global thermosphere-ionosphere system, but the time-scales of the influence, as well as the dynamics and chemistry, are poorly understood. As another example, the large-scale thermospheric and ionospheric structures are known to cause small-scale events, such as ionospheric scintillation.

4. The changing magnetic field of the Earth. In order to understand the long-term stability of the magnetosphere, it is important to understand the processes that create the main field.

Benefits to Society

Satellites cost millions of dollars to build. Satellites can be damaged or even destroyed by a number of phenomena that occur in the near Earth space environment, such as radiation erosion of electronics, charging of space craft and subsequent arching, which may occur, and the ejection of the space-craft from the magnetic stability region of space.

Understanding of the near-Earth space environment is quite important because of the increasing number of space assets, the significant cost of launching satellites, and the monetary investment in developing and building these satellites

8.2 Short-Term Priorities

8.2.1 Thermosphere and Ionosphere

Prediction of ionospheric scintillation formation.

Ionospheric scintillation occurs when the ionosphere becomes unstable to the Rahliegh-Taylor instability. The main mechanism for this instability is thought to be understood, but the prediction of exactly when and where this occurs is not perfected. Scintillation basically creates giant plumes of ionospheric density at very high altitudes (~ 300 km) *Tsunoda* [1988]. These enhanced densities cause significant radar back scatter and make high-frequency communication nearly impossible. It is therefore important to understand when and where scintillation will occur.

Influence of gravity waves on ionospheric structure.

It is known that mesospheric gravity waves propagate up to the thermosphere, and that these waves can cause significant changes in the chemistry and dynamics of the lower thermosphere [e.g. *Fuller-Rowell* [1994]; *Hickey et al.* [1992]]. Gravity waves can therefore affect the ionosphere also through chemistry and ion neutral drag forces. When the gross structure of gravity waves is well understood, the exact global structure is not. In addition, the quantification of the effects upon the ionospheric structure and dynamics needs to be completed. This will most likely be done through the use of global thermosphere and ionosphere models which have a more accurate gravity wave structure included. An effort is underway at the National Center for Atmospheric Research to build a whole atmosphere community climate model (WACCM) which will model from the troposphere through the thermosphere. This model will include all of the large-scale waves which propagate upward into the thermosphere and ionosphere.

Dynamics and chemistry due to Joule heating.

With the launch of the TIMED satellite, quantitative understanding of thermospheric Joule heating has become possible at small spatial scales. Joule heating is caused by ions colliding with neutrals, which are moving with a different velocity. This heating occurs in regions in which there is significant ion density and differential motion [e.g. *Ahn et al.* [1983]; *Foster et al.* [1983]; *Chun et al.* [1999]]. The heating causes the thermosphere to expand and waves to be launch from the polar region towards the equator. The expansion of the thermosphere can increase satellite drag and can cause ionospheric outflow, which adds particles to the ring current pop-

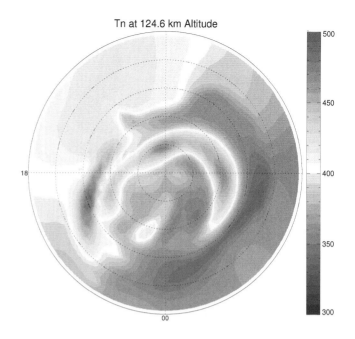

Tn at 124.6 km Altitude

Figure 8.1: This figure shows the thermospheric temperature structure at 125 km altitude above the North polar region. The top of the plot is towards the Sun while the left side is dusk. Each circle is $10°$, while the outer circle is $40°$ geographic latitude. The enhanced temperatures are caused by Joule heating.

ulation. The propagating waves are typically called traveling ionospheric disturbances, and cause problems with high frequency radars and communication devices. Figure 8.1 shows the thermospheric temperature structure during a storm which took place on May 4, 1998.

Investigation of small scale structures in global models.

A number of global models of the thermosphere – ionosphere system exist today. One of the most important models of the thermospheric region is MSIS [*Hedin*, 1983, 1987, 1991]. MSIS is an empirical model which relates the neutral densities and temperature to $F10.7$ and Ap. It uses a spherical harmonic fit of many different satellite and remote observations. It is considered one of the more important models because of the amount of data which has gone into the model, and therefore it is felt that it represents the average conditions for the given inputs. MSIS has been improved by adding time dependence, which allows storm periods to be modeled more accurately.

The international reference ionosphere (IRI) is similar to MSIS, but for the ionosphere [*Rawer et al.*, 1978; *Bilitza*, 2001]. It models the ionospheric density and temperatures for all latitudes and local times, but does not include any auroral precipitation, which makes the model incorrect in the auroral regions for medium and strong auroral activity.

Other types of thermosphere or ionosphere models are the

first principle models which model the different regions by determining the density, momentum, and energy of the fluid self-consistently. The T-GCM (TGCM, TIGCM, TIEGCM, TIMEGCM) series of models are examples of such models which were created at the National Center for Atmospheric Research. These models solve for mass mixing ratios of the neutral major species O_2, N_2, and O and the minor species $N(^2D)$, $N(^4S)$, NO, He, and Ar. They are full 3-dimensional codes with 5° latitude by 5° longitude by 0.5 scale height altitude cells [*Dickinson et al.*, 1981, 1984; *Roble et al.*, 1988; *Roble and Ridley*, 1987; *Richmond et al.*, 1992; *Roble and Ridley*, 1994]. A different first principles model is the Coupled Thermosphere Ionosphere Model (CTIM) [*Fuller-Rowell and Rees*, 1980, 1983; *Rees and Fuller-Rowell*, 1988; *Rees. and Fuller-Rowell*, 1990].

Wang et al. [1999] and *Wang et al.* [2001] extended the TIGCM to include a high resolution nested grid to resolve features which are less than 5° in latitude or longitude. They typically use the nested grid within the auroral region. The maximum resolution of the nested grid is 1.3°.

In the next decade, it is expected that the global thermosphere - ionosphere models will attain much higher resolution than current models. This will allow for the investigation of how intense small-scale phenomena effect the regional and global environment. For example, auroral arcs are known to be only a few tens of kilometers across down to meters across and can be associated with large electric fields and particle precipitation fluxes. This causes intense heating in a very localized environment, which may spread through diffusion or may travel through waves. Current models can not come close to examining these types of effects.

Data assimilation.

There are currently only a small number of groups who are actively attempting to perform data assimilation within the thermosphere and ionosphere. The assimilative mapping of ionospheric electrodynamics (AMIE) technique ingests electrodynamic data to determine the electric field, particle precipitation, and current structure of the high-latitude ionosphere [e.g., *Richmond and Kamide* [1988]; *Richmond* [1992]; *Ridley et al.* [1997, 1998]]. While this technique is quite useful, it only works on a 2-dimensional grid in the high-latitude region, and it only produces maps of electrodynamic quantities and not fundamental quantities such as densities, velocities, and temperatures.

One group has started to incorporate measurements of total electron content into a 3-dimensional model of the ionosphere [see http://gaim.cass.usu.edu/index.html]. This is being carried out by using data assimilation techniques developed in weather forecasting (i.e. Kalman filtering) and applying them to global simulations of the ionosphere. In the next decade, it is expected that a number of other groups will begin to examine using Kalman filtering within thermosphere, ionosphere, and magnetosphere codes.

Influence of Earth's ionosphere and magnetosphere on the lower atmosphere.

Our understanding of how the magnetosphere influences the mesosphere, stratosphere, and troposphere is quite limited. It is known that there is relatively little influence over the period of days to weeks, but some correlation has been shown to occur over decadal time periods. The most common example of this is the Maunder minimum (1650-1710). This is a period of time in which there were very few Sun spots, so the magnetic activity in the near-Earth space environment was quite low. During this time, the Earth underwent a mini ice age, which lasted about the same amount of time as the low activity. The problem with this correlation is that the solar luminosity is also tied to Sun spots, so during this period, the Sun may not have been heating Earth quite as much as it normally does.

There is currently an effort underway to couple lower atmosphere models to the T-GCM series of thermosphere-ionosphere models. This effort is called the Whole Atmosphere Community Climate Model (or WACCM). When this model is completed and applied to a large number of events or a large time period, research may begin to determine how much coupling actually occurs between these regions of our atmosphere.

Prediction of ionospheric current structures.

The first order global ionospheric field-aligned current structure was described many years ago [*Iijima and Potemra*, 1976]. It is known that within this global structure there are many other smaller structures. These small-scale structures can be less that 100 km in latitudinal width and can move around quite quickly. While the exact temporal and spatial structure is difficult to study, the dynamics which control this may be understood, and specifications of the types of structures (e.g. stable location versus highly variable location) are important to learn about.

Space Weather.

There are a number of research groups which are starting to run models in real-time, but have not fully validated them, nor do they insure that they will run automatically 24 hours a day, 7 days a week. In the next decade, it is expected that a number of researchers will push their models to become operational codes. This will allow space weather monitoring stations, such as the Space Environment Center, to use a variety of different codes for determining the conditions in the near-Earth space environment.

For tropospheric weather forecasting, a number of models are run and are combined together to create an ensemble average forecast. It will most likely be a number of years before this type of averaging can be done within the thermospheric - ionospheric community, but we expect there will be a lot of progress within this field.

8.2.2 Magnetosphere

Stability and dynamics of the magnetospheric tail

Magnetospheric substorms are global events, which produce aurora near the Earth's magnetic poles [as first described by *Akasofu* [1964]]. Earth's magnetospheric tail can grow to be very long and thin (i.e. 1/2 Earth Radius) in Z. This thin tail has a large gradient in magnetic field orientation, which drives a strong cross-tail current (or vise-versa, depending on your view point). At some point, the cross tail current becomes disrupted, and the current is diverted from the tail into the ionosphere near the poles. This produces large auroral displays near the current regions. Figure 8.2 shows the magnetosphere with a very long tail (possibly a highly energized state), and a short tail (a low energy state).

The exact mechanism, which causes the disruption of the cross-tail Current, is unknown. We list a few models of substorms here: the near-Earth neutral line model [e.g. *Hones* [1979]], the boundary layer model [*Rostoker and Eastman*, 1987], the thermal catastrophe model [*Goertz and Smith*, 1989], the current disruption model [*Chao et al.*, 1977; *Lui*, 1996], the Magnetosphere – Ionosphere coupling model [*Kan*, 1993], and the interplanetary magnetic field triggered substorm model [*Lyons*, 1995]. In addition, the current may not disrupt over the whole tail; it may occur in localized regions [e.g. *Lyons et al.* [2003]], which complicates issues significantly since instrumentation, which measures these phenomena, is quite sparse. So, even identifying true global-scale events is problematic.

The search for reconnection continues.

One of the most important processes for transferring energy from the solar environment to Earth's magnetosphere is reconnection [*Dungey*, 1961]. This process changes magnetic energy into particle energy. The understanding of this process is one of the most fundamental for understanding how the magnetosphere interacts with the solar environment. It is not well understood. There are still controversies about whether reconnection has to occur for antiparrellel magnetic field lines or whether component merging can take place. In the magnetospheric tail, the reconnection process and its relations to current disruption is even less understood (as discussed above).

With the launch of the Cluster satellites, a number of researchers have started to search for magnetospheric reconnection sites [e.g. *Phan et al.* [2003]]. Recently, researchers have shown evidence that the reconnection site has moved over the cluster of satellites. Studies with the Polar satellite have also shown evidence of crossing the reconnection region. More of these types of events will be found in the near future, and will help us to address some fundamental issues in reconnection theory.

Data assimilation.

Data assimilation within the magnetosphere is almost nonexistent. There are a number of reasons for this: (1) lack of data to be assimilated; (2) the strong nonlinear processes within the magnetosphere; and (3) the strong spatial discontinuities, which exist because of the magnetic field. It

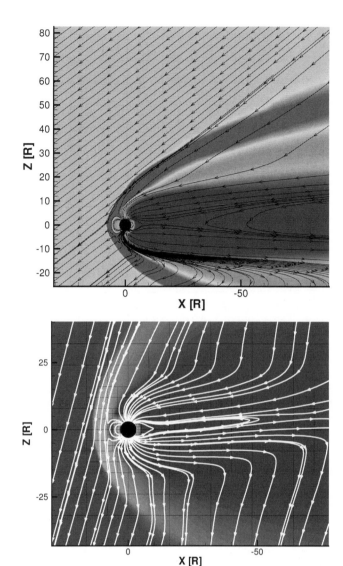

Figure 8.2: These two figures show examples of how the magnetospheric tail can be quite short (top) or long (bottom), depending on the interaction of the magnetosphere with the solar wind and IMF. The color indicates the density (different scales), while the line traces are the magnetic field lines. The black circle represents the inner boundary on the magnetospheric 3-D simulation. The Sun is off to the left.

is expected that over the next decade researchers will begin to investigate the use of data within global and localized simulations of the magnetosphere.

Radiation belts.

The radiation belts are bands of very high-energy particles, which are at relatively low altitudes (within approximately 2 Earth radii). Because of the altitudes of the radiation belts, many orbital planes cut through them. The exposure of satellite electronics to these particles is highly degrading, and limits the lifetime of the satellite severely. It may would be useful and important if in the future we could more accurately predict the temporal and spatial specification of the radiation belts [e.g. *Elkington et al.* [2003]]. This may allow satellite operators to extend the life of their satellites.

Influence of the thermosphere and ionosphere on the magnetospheric dynamics.

Recently, researchers have started to examine more closely how the ionosphere influences the magnetosphere. There are a number of areas within this field which have been examined on a small scale before. For example, it is well known that the magnetospheric ring current is partially composed of oxygen atoms, which originate from the ionosphere. Therefore it can be said that ionospheric outflow plays a role in the amount of energy that can be fed into the ring current.

Another example is the control of the thermospheric neutral winds over the ionospheric and magnetospheric dynamics. *Thayer and Vickrey* [1992]; *Sánchez et al.* [1998]; *Thayer* [1998] make use of the Sondrestrom incoherent radar to determine how the thermospheric neutral winds effect the energy flow from the magnetosphere to the ionosphere and back. *Deng et al.* [1991, 1993]; *Lu et al.* [1995] use a global thermosphere - ionosphere model to determine the effect of the neutral winds on the ionosphere. The models presented by *Peymirat et al.* [1998], *Peymirat et al.* [2002], and *Ridley et al.* [2003b] show results of a coupled TIEGCM different magnetospheric models. These studies start to quantify the effects of the neutral winds on the global magnetospheric convection and configuration.

Fedder and Lyon [1987] showed that the magnetosphere is neither a current nor a voltage generator by using a global MHD code by varying a constant ionospheric conductance to determine whether the magnetosphere would self-consistently increase the field-aligned currents or not. *Raeder et al.* [1996] showed that the ionospheric conductance specification is crucial in determining the temporal history of the magnetospheric dynamics. They showed that by changing the conductance in the ionosphere, they could make the magnetosphere go into either a steady convection state or storage-unloading state. *Raeder et al.* [2001] further describes how the ionospheric conductance can determine the timing and strength of auroral substorms through the use of a limited parametric study. The study by *Ridley et al.* [2003a] begins to examine the effect of the different ionospheric conductance patterns on the global state of the magnetosphere.

One of the main issues with these types of studies is the lack of diverse measurement conditions to prove the cause and effect statistically. This must change in order to determine how much of an effect the thermosphere - ionosphere system has on the magnetosphere.

Energy flow through the magnetosphere.

Statistical studies have shown that embedded regions of southward interplanetary magnetic fields with intensity $B_z < -10nT$ and duration of greater than three hours in the solar wind produce major magnetic storms at Earth with Dst dip-

ping below -100 nT [*Gonzalez and Tsurutani*, 1987]. The physical mechanism underlying the energy transfer from the solar wind to the magnetosphere is reconnection at the dayside magnetopause. These regions of strong southward IMF are termed "geoeffective" which is widely interpreted to mean that they create large geomagnetic storms. The intensity of the ring current enhancement is a primary measure of the magnitude of the geomagnetic storm. However, the way in which energy is distributed throughout the global magnetosphere in association with various types of solar wind transients is not known. Recent studies indicate that energy flow through the inner magnetosphere and into the ring current reservoir may account for as little as 20% of the total energy dissipated during a major magnetic storm [*Kozyra et al.*, 1998; *Weiss et al.*, 1992]. If this is the case, then a large part of the magnetic storm energy is loaded into the magnetotail lobes, lost downtail as plasmoids and dissipated in the auroral ionosphere. It is possible that the majority of the energy input during magnetic storms is contained in the highly disturbed magnetotail configuration and dissipated in frequent and intense substorms. In fact, during the main phase of magnetic storms, magnetic field depolarizations and substorm injection signatures are so frequent that identifying individual substorms is a nearly impossible task [c.f., *Kamide et al.* [1998]]. Night side growth phase and injection signatures sometimes occur simultaneously, separated by only a few hours in local time. This high number of substorm recurrences must lead to elevated stormtime plasma sheet temperatures and large amounts of energy pumped into the auroral ionosphere.

Over the next decade, a major focus of magnetospheric research is going to be attempting to quantify how much energy is entering the magnetospheric system and partitioning the loss of that energy into the different regions of the magnetosphere. Specifically, the IMAGE mission is examining the ring current dynamics, while the TIMED mission is examining the energy deposition into the ionosphere. One of the roles of the Cluster mission is to examine the dynamics of the plasma/current sheet, which will help to determine how much energy is flowing from the tail towards the Earth and how much is being lost down tail.

8.2.3 The Sun and the Heliosphere

Helioseismology

The science studying wave oscillations in the Sun is called helioseismology. One can view the physical processes involved in the same way that seismologists learn about Earth's interior by monitoring waves caused by earthquakes [*Chaplin et al.*, 2002; *Dzhalilov et al.*, 2002]. Temperature, composition, and motions deep in the Sun influence the oscillation periods and yield insights into conditions in the solar interior. Because it is impossible to send satellites to sample the solar interior, helioseismology offers one of the only methods of measuring the dynamics of the solar interior. Most of the

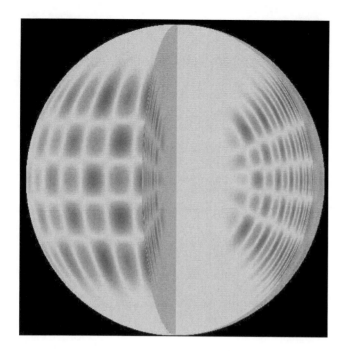

Figure 8.3: This image was generated by the computer to represent an acoustic wave resonating in the interior of the Sun. Here, the radial order is $n = 14$, angular degree is $l = 20$, and the angular order is $m = 16$. Red and blue show element displacements of opposite sign. The frequency of this mode determined from the MDI data is $2935.88 +/- 0.2 \, \mu Hz$.

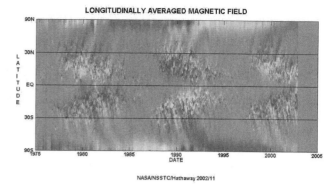

Figure 8.4: A magnetic field butterfly diagram. This image shows how the magnetic field near the solar surface changes as a function of time. Equatorward propagation of the magnetic field is evident.

physics of the Sun derives from its interior, so understanding the interior is of primary importance. That makes helioseismology very important also. It is expected that techniques for measuring the waves on the solar surface will improve over the coming decade, so that determining the interior dynamics will be much easier.

The Solar Dynamo

It is widely believed that the Sun's magnetic field is generated by a magnetic dynamo within the Sun. The fact that the Sun's magnetic field changes dramatically over the course of just a few years, and the fact that it changes in a cyclical manner indicates that the magnetic field continues to be generated within the Sun [*Solanki et al.*, 2002]. The solar dynamo is responsible for several phenomena: 1) the 11-year period of the sunspot cycle, 2) the equator-ward drift of the active latitude, 3) Hale's polarity law and the 22-year magnetic cycle, 4) Joy's law for the observed tilt of sunspot groups and, 5) the reversal of the polar magnetic fields near the time of cycle maximum.

The reversing of the Sun's dynamo causes significant activity to occur on the solar surface every 11 years. When the Sun's magnetic field is close to dipolar, there is very little activity, but when the field is strongly tilted, there are many active regions and coronal mass ejections (see Figure 8.4). This is of primary importance to Earth because every 11 years,

when the dipole is highly tilted and the Sun is quite active, large events are launched towards the Earth, and dramatic magnetic storms occur. While there are a number of theories of why the dipole flips every 11 years, there is little agreement. This will continue to be a large topic in space research until it is solved.

Coronal mass ejections

Coronal mass ejections (or CMEs) are huge bubbles of gas threaded with magnetic field lines that are ejected from the Sun over the course of several hours [*Subramanian and Dere*, 2001]. Although the Sun's corona has been observed during total eclipses of the Sun for thousands of years, the existence of coronal mass ejections was unrealized until the space age. The earliest evidence of these dynamical events came from observations made with a coronagraph on the 7th Orbiting Solar Observatory (OSO 7) from 1971 to 1973. A coronagraph produces an artificial eclipse of the Sun by placing an "occulting disk" over the image of the Sun. During a natural eclipse of the Sun the corona is only visible for a few minutes at most, too short a period of time to notice any changes in coronal features. With ground based coronagraphs, only the innermost corona is visible above the brightness of the sky. From space the corona is visible out to large distances from the Sun and can be viewed continuously.

CMEs disrupt the flow of the solar wind and produce disturbances that strike the Earth with sometimes catastrophic results. The Large Angle and Spectrometric Coronagraph (LASCO) on the Solar and Heliospheric Observatory (SOHO) has observed a large number of CMEs (e.g. Figure 8.5). Understanding how CMEs occur and what triggers them to erupt is very important to being able to predict when there will be activity at Earth. It is similar to the substorm research, which is being conducted in Earth's magnetosphere - both are explosive events, which are possibly triggered. The triggering mechanism is still unknown for both [*Priest and Forbes*, 2002; *Dikpati et al.*, 2002].

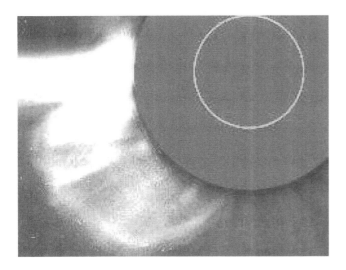

Figure 8.5: The CME event of April 7th, 1997. It produced a "halo event" in which the entire Sun appeared to be surrounded by the CME. Halo events are produced by CMEs that are directed toward (or away from) the Earth. As they loom larger and larger they appear to envelope the Sun itself.

Solar energetic particles

Starting in 1942, Geiger counters and other detectors set up to monitor cosmic rays have occasionally seen sudden increases in the intensity of the radiation associated with outbursts on the Sun, mostly with visible flares [*Gopalswamy et al.*, 2002]. The cosmic ray intensity returns to normal within minutes or hours as the acceleration process ends, and as accelerated ions disperse throughout interplanetary space [*Aschwanden*, 2002].

In many events, the Sun emits enormous numbers of lower-energy ions with no more than tens of Mev (=millions electron volts). The Earth's magnetic field diverts them to the vicinity of the magnetic poles where they may temporarily smother the ionosphere and interfere with radio communications. Such "polar cap blackouts" used to bother US military radar installations, which scanned the polar cap for hostile missiles.

Should we be concerned about solar energetic radiation? On Earth, we are safe, protected by the thick layer of our atmosphere, equivalent to 10 meters (32 feet) of water or about 4 meters of concrete. Astronauts on a space station orbiting near the Earth's equator are protected by the Earth's magnetic field but not enough for some solar particle event. In those cases, they are required to return to their spacecraft. Thus, predicting these events is important to astronauts working outside their space vehicles. Also, astronauts on their way to Mars, separated from space by just a thin metal shell, are quite vulnerable. Luckily, life threatening radiation events are rare, especially during low years of the sunspot cycle. Still, as the large particle events of August 1972 have shown, some danger remains even then.

Researchers are just starting to incorporate solar energetic particles into global heliosphere models. Having CMEs erupting into the global heliosphere will create large reconnection sites and strong shock structures. In the future, it is expected that researchers will better understand the acceleration mechanisms for these high energy particles.

Propagation of structures through interplanetary space

When CMEs are observed near the Sun, it is unknown whether they will encounter the Earth or not. If they are suspected to encounter the Earth, the exact time of arrival and intensity of the CME and the pile up of material in front of it are unknown until these structures are measured by satellites located approximately 250 Earth Radii upstream of the Earth. CMEs typically take 24-72 hours to arrive at the Earth. Figure 8.6 shows an example of a 3-D simulation of a CME encountering the Earth.

The propagation of the CME through interplanetary space is crucial in determining how much of an effect it will have on the Earth system. If it is a fast CME moving through a slow medium, there will be a huge pile-up in front of it, and this pile-up may cause as much of an effect as the CME itself. In addition, the slow medium will cause the CME to slow down, and it will take longer to reach the Earth.

Sophisticated models of the heliosphere and propagation of CMEs through the heliosphere are really in their infancy. Since the propagation is so important in determining when the CME will arrive and how strong it will be, significant resources are being put towards the improvement of these models.

8.2.4 Ground-based Measurements

More observatories

In order to examine the global structure of the thermospheric, ionospheric, and magnetospheric dynamics, it is crucial to have a wide variety of data sources covering as much of the globe as possible. IAGA research covers a huge region of space, but has an extremely small number of data sources. It is surprising that any progress can be made at all. At any given time, there are less than 200 magnetometers running. This means that each magnetometer, on average, has to cover 2.55 million km^2. What it *really* means is that some regions have relatively poor coverage, and other regions have awful coverage. One of the worst parts is that the coverage is not based on economy of the country at all. So, for example, the United States has very poor data coverage considering the amount of money that is spent each year on research.

In order for progress to be made at any pace above what is being done today, more observatories are needed. While some researchers understand this and are proposing for chains of observatories, there are not enough. In addition, the funding agencies have to realize that ground-based observatories are very inexpensive ways to help scientists better understand our environment.

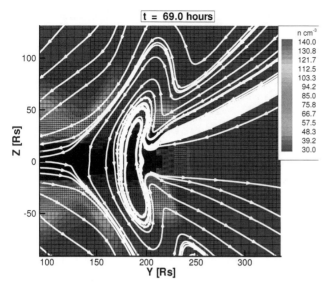

Figure 8.6: This plot shows the distortion of a CME which started as a tear-drop shape near the Sun. The line traces are the magnetic field lines of the Sun. Here it is close to 1 AU, and it has been significantly slowed in the solar current sheet. Above and below the current sheet, the CME is moving faster, so the shape is distorted. This figure is from a global 3-D simuation.

Creation of centralized databases

One of the main difficulties in conducting large-scale or global studies in the magnetospheric and ionospheric domain is the lack of a consistent and centralized database. In order to carry out this type of research, which is truly the future of geophysics, databases must either be cross-linked, or centralized. Data formats must be created which allow users to easily process and examine the data. All data providers (or the centralized data storage facility) must buy into the specific format(s), to allow users to use data from multiple chains. Without these capabilities, advancement in geophysics is going to come at a slow pace. A number of researchers are working on these problems and hopefully progress will be made in this area.

8.3   Recommendations for the IAGA

In order to accurately predict the near-Earth space environment with more than a one hour lead time, a better understanding of the solar environment must be accomplished. The prediction capabilities can be extended up to approximately 3 days if better models of the interplanetary medium are created and more accurate specification of coronal mass ejections and other large-scale phenomena near the solar surface are accomplished.

In addition to improving the lead-time of the interplanetary conditions, a better understanding of the magnetosphere-ionosphere system must be accomplished. This will most likely be achieved by:

- Use of coupled models of the magnetospheric systems. A number of groups are coupling large-scale magnetohydrodynamic models of the magnetosphere with particle codes which model the inner magnetosphere and thermosphere-ionosphere models. These coupled models will allow researchers to accurately model the complete magnetospheric-ionospheric system and carry out numerical experiments within this system to develop a much better understanding of how the near-Earth space environment reacts to different interplanetary medium.

- Combining magnetospheric data with global models to put the data into context. While a number of researchers attempt to do this, it is typically in a model validation type of study. In the next few years, it is expected that more data analysts will turn to modelers to put measurements into context. This combination of model and data will help to determine the first order effects as well as showing how any subgrid-scale phenomena effect the structure of the whole system. There is a lot of work to be done in this area, but it is one of the most crucial areas of research today.

- More measurements. The magnetosphere and ionosphere are data poor. The community must be more vocal in setting up new instrumentation and getting new satellites launched. Microsatellite technology applied tens to hundreds of times is needed to cover as much of the magnetospheric system as possible. In addition, using innovative remote-sensing systems, such as those on the IMAGE satellite, can help to examine many magnetospheric phenomena at once and in a larger context.

Most of these improvements can be made because of the increase in computing power over the past 30 years, as well as new technology. Space science will continue to use these new tools to make discoveries.

8.4   Conclusions

It is an exciting time to be a part of IAGA. Many of the basic properties of the system seem to be well understood. This has created the demand for predictive models of the system - taking the physics and turning it into engineering. Also, because many of the basic properties are understood, we now have the ability to study the non-linear and cross-scale coupling that is thought to make the system so dynamic and interesting.

8.5   Acknowledgments

The author would like to thank those who responded to the survey which was conducted in the summer of 2002. I would also like to thank Mike Liemohn, Janet Kozyra, Laura Peticolis, and Ilia Roussev for their helpful discussions on the future of IAGA and reading this document.

Bibliography

Ahn, B.-H., S.-I. Akasofu, and Y. Kamide, The joule heating production rate and particle energy injection rate as a function of the geomagnetic indices AE and AL, *J. Geophys. Res.*, *88*, 6275, 1983.

Akasofu, S.-I., The development of the auroral substorm, *Planet. Space Sci.*, *101*, 273, 1964.

Aschwanden, M. J., Particle acceleration and kinematics in solar flares - A Synthesis of Recent Observations and Theoretical Concepts (Invited Review), *Space Science Reviews*, *101*, 1, 2002.

Bilitza, D., International reference ionosphere 2000, *Radio Science*, *36*, 261, 2001.

Chao, J.K., J.R. Kan, A.T.Y. Lui, and S.-I. Akasofu, A model for thinning of the plasma sheet, *Planet. Space Sci.*, *25*, 703, 1977.

Chaplin, W. J., Y. Elsworth, G. R. Isaak, B. A. Miller, and R. New, On the measurement precision of solar p-mode eigenfrequencies, *Monthly Not. of the Royal Astron. Soc.*, *330*, 731, 2002.

Chun, F. K., D. J. Knipp, M. G. McHarg, G. Lu, B. A. Emery, S. Vennerstrom, and O. A. Troshichev, Polar cap index as a proxy for hemispheric joule heating, *Geophys. Res. Lett.*, *26*, 1101, 1999.

Deng, W., T.L. Killeen, A.G. Burns, and R.G. Roble, The flywheel effect: Ionospheric currents after a geomagnetic storm, *Geophys. Res. Lett.*, *18*, 1845, 1991.

Deng, W., T.L. Killeen, A.G. Burns, R.G. Roble, J.A. Slavin, and L.E. Wharton, The effects of neutral inertia on ionospheric currents in the high-latitude thermosphere following a geomagnetic storm, *J. Geophys. Res.*, *98*, 7775, 1993.

Dickinson, R.E., E.C. Ridley, and R.G. Roble, A three-dimensional, time-dependent general circulation model of the thermosphere, *J. Geophys. Res.*, *86*, 1499, 1981.

Dickinson, R.E., E.C. Ridley, and R.G. Roble, Thermospheric general circulation with coupled dynamics and composition, *J. Atmos. Sci.*, *41*, 205, 1984.

Dikpati, M., T. Corbard, M. J. Thompson, and P. A. Gilman, Flux Transport Solar Dynamos with Near-Surface Radial Shear, *Astrophys. J. Lett.*, *575*, L41, 2002.

Dungey, J.W., Interplanetary magnetic field and the auroral zones, *Phys. Rev. Lett.*, *93*, 47, 1961.

Dzhalilov, N. S., J. Staude, and V. N. Oraevsky, Eigenoscillations of the differentially rotating Sun. I. 22-year, 4000-year, and quasi-biennial modes, *Astron. Astrophys.*, *384*, 282, 2002.

Elkington, S.R., M.K. Hudson, and A.A. Chan, Resonant acceleration and diffusion of outer zone electrons in an asymmetric geomagnetic field, *J. Geophys. Res.*, 2003.

Fedder, J.A., and J.G. Lyon, The solar wind – magnetosphere – ionosphere current – voltage relationship, *Geophys. Res. Lett.*, *14*, 880, 1987.

Foster, J.C., F.-P. St. Maurice, and V.J. Abreu, Joule heating at high latitudes, *J. Geophys. Res.*, *88*, 4885, 1983.

Fuller-Rowell, T.J., and D. Rees, A three-dimensional, time-dependent, global model of the thermosphere, *J. Atmos. Sci.*, *37*, 2545, 1980.

Fuller-Rowell, T.J., and D. Rees, Derivation of a conservative equation for mean molecular weight for a two constituent gas within a three-dimensional, time-dependent model of the thermosphere, *Planet. Space Sci.*, *31*, 1209, 1983.

Fuller-Rowell, T.J., The impact of gravity waves on nitric oxide in the lower thermosphere, *J. Atmos. Terr. Phys.*, *56*, 1387, 1994.

Goertz, C.K., and R.A. Smith, Thermal catastrophe model of substorms, *J. Geophys. Res.*, *94*, 6581, 1989.

Gonzalez, W.D., and B.T. Tsurutani, Criteria of interplanetary parameters causing intense magnetic storms ($D_{st} < -100nT$), *Planet Space Sci.*, *35*, 1101, 1987.

Gopalswamy, N., S. Yashiro, G. Michałek, M. L. Kaiser, R. A. Howard, D. V. Reames, R. Leske, and T. von Rosenvinge, Interacting Coronal Mass Ejections and Solar Energetic Particles, *Astrophys. J. Lett.*, *572*, L103, 2002.

Hedin, A.E., A revised thermospheric model based on mass spectrometer and incoherent scatter data: MSIS-83, *J. Geophys. Res.*, *88*, 10170, 1983.

Hedin, A.E., MSIS-86 thermospheric model, *J. Geophys. Res.*, *92*, 4649, 1987.

Hedin, A.E., Extension of the MSIS thermosphere model into the middle and lower atmosphere, *J. Geophys. Res.*, *96*, 1159, 1991.

Hickey, M.P., G. Schubert, and R.L. Walterscheid, Seasonal and latitudinal variations of gravity wave-driven fluctuations in oh nightglow, *J. Geophys. Res.*, *97*, 14,911, 1992.

Hones, E.W. Jr., Transient phenomena in the magnetotail and their relation to substorms, *Space Sci. Rev.*, *23*, 393, 1979.

Iijima, T., and T. A. Potemra, The amplitude distribution of field-aligned currents at northern high latitudes observed by Triad, *J. Geophys. Res.*, *81*, 2165, 1976.

Kamide, Y., W. Baumjohann, I.A. Daglis, W.D. Gonzalez, M. Grande, J.A. Joselyn, R.L. McPherron, J.L. Phillips, E.G.D. Reeves, G. Rostoker, A.S. Sharma, H.J. Singer, B.T. Tsurutani, and V. M. Vasyliunas, Current understanding of magnetic storms: Storm-substorm relationships, *J. Geophys. Res.*, *103*, 17,705, 1998.

Kan, J.R., A global Magnetosphere – Ionosphere coupling model of substorms, *J. Geophys. Res.*, *98*, 17,263, 1993.

Kozyra, J. U., V. K. Jordanova, J. E. Borovsky, M. F. Thomsen, D. J. Knipp, D. S. Evans, D. J. McComas, and T. E. Cayton, Effects of a high density plasma sheet on ring current development during the November 2-6, 1993 magnetic storm, *J. Geophys. Res.*, *103*, 26,285, 1998.

Lu, G., A.D. Richmond, B.A. Emery, and R.G. Roble, Magnetosphere-ionosphere-thermosphere coupling: Effect of neutral winds on energy transfer and field-aligned current, *J. Geophys. Res.*, *100*, 19,643, 1995.

Lui, A.T.Y., Current disruption in the earth's manetosphere: Observations and models, *J. Geophys. Res.*, *101*, 13,067, 1996.

Lyons, L. R., E. Zesta, Y. Xu, E. R. Sánchez, J. C. Samson, G. D. Reeves, J. M. Ruohoniemi, and J. B. Sigwarth, Auroral poleward boundary intensifications and tail bursty flows: A manifestation of a large-scale ULF oscillation?, *J. Geophys. Res.*, 2003.

Lyons, L.R., A new theory for magnetospheric substorms, *J. Geophys. Res.*, *100*, 19,069, 1995.

Peymirat, C., A.D. Richmond, B.A. Emery, and R.G. Roble, A magnetosphere-thermosphere-ionosphere electrodynamics general circulation model, *J. Geophys. Res.*, *103*, 17467, 1998.

Peymirat, C., A.D. Richmond, and R.G. Roble, Neutral wind influence on the electrodynamic coupling between the ionosphere and the magnetosphere, *J. Geophys. Res.*, *107*, 1029, 2002.

Phan, T., H. U. Frey, S. Frey, L. Peticolas, S. Fuselier, C. Carlson, H. Rème, J.-M. Bosqued, A. Balogh, M. Dunlop, L. Kistler, C. Mouikis, I. Dandouras, J.-A. Sauvaud, S. Mende, J. McFadden, G. Parks, E. Moebius, B. Klecker, G. Paschmann, M. Fujimoto, S. Petrinec, M. F. Marcucci, A. Korth, and R. Lundin, Simultaneous Cluster and IMAGE observations of cusp reconnection and auroral proton spot for northward IMF, *J. Geophys. Res.*, 2003.

Priest, E. R., and T. G. Forbes, The magnetic nature of solar flares, *Astron. Astrophys. Rev.*, *10*, 313, 2002.

Raeder, J., J. Berchem, and M. Ashour-Abdalla, The importance of small scale processes in global MHD simulations: Some numerical experiments, In Chang, T., and J. R. Jasperse, editors, *The Physics of Space Plasmas*, volume 14, page 403, Cambridge, Mass., 1996. MIT Cent. for Theoret. Geo/Cosmo Plasma Phys.

Raeder, J., R.L. McPherron, L.A. Frank, S. Kokubun, G. Lu, T. Mukai, W.R. Paterson, J.B. Sigwarth, H.J. Singer, and J.A. Slavin, Global simulation of the Geospace Environment Modeling substorm challenge event, *J. Geophys. Res.*, *106*, 281, 2001.

Rawer, K., D. Bilitza, and S. Ramakrishnan, Goals and status of the international reference ionosphere, *Rev. Geophys.*, *16*, 177, 1978.

Rees, D., and T.J. Fuller-Rowell, Understanding the transport of atomic oxygen in the thermosphere using a numerical global thermospheric model, *Plan. Space Sci.*, *36*, 935, 1988.

Rees., D., and T.J. Fuller-Rowell, Numerical simulations of the seasonal/latitudinal variations of atomic oxgen and nitric oxide in the lower thermosphere and mesosphere, *Adv. Space Res.*, *10(6)*, 83–102, 1990.

Richmond, A.D., and Y. Kamide, Mapping electrodynamic features of the high-latitude ionosphere from localized observations: Technique, *J. Geophys. Res.*, *93*, 5741, 1988.

Richmond, A.D., E.C. Ridley, and R.G. Roble, A thermosphere/ionosphere general circulation model with coupled electrodynamics, *Geophys. Res. Lett.*, *19*, 369, 1992.

Richmond, A.D., Assimilative mapping of ionospheric electrodynamics, *Adv. Space Res.*, *12*, 59, 1992.

Ridley, A.J., C.R. Clauer, G. Lu, and V.O. Papitashvili, Ionospheric convection during nonsteady interplanetary magnetic field conditions, *J. Geophys. Res.*, *102*, 14,563, 1997.

Ridley, A.J., C.R. Clauer, G. Lu, and V.O. Papitashvili, A statistical study of the ionospheric convection response to changing interplanetary magnetic field conditions using the assimilative mapping of ionospheric electrodynamics technique, *J. Geophys. Res.*, *103*, 4023, 1998.

Ridley, A.J., T.I. Gombosi, and D.L. De Zeeuw, Ionospheric control of the magnetospheric configuration (1): Conductance, *Annales Geophys.*, 2003a, In Press.

Ridley, A.J., T.I. Gombosi, D.L. De Zeeuw, C.R. Clauer, and A.D. Richmond, Ionospheric control of the magnetospheric configuration: Neutral winds, *J. Geophys. Res.*, 2003b, In Press.

Roble, R.G., and E.C. Ridley, An auroral model for the NCAR thermospheric general circulation model (TGCM), *Ann. Geophys.*, *5A*, 369, 1987.

Roble, R.G., and E.C. Ridley, A thermosphere-ionosphere-mesosphere-electrodynamics general circulation model (time-gcm): Equinox solar cycle minimum simulations (300-500km), *Geophys. Res., Lett.*, *21*, 417, 1994.

Roble, R.G., E.C. Ridley, A.D. Richmond, and R.E. Dickinson, A coupled thermosphere/ionosphere general circulation model, *Geophys. Res. Lett.*, *15*, 1325, 1988.

Rostoker, G., and T.E. Eastman, A boundary layer model for magnetospheric substorms, *J. Geophys. Res.*, *92*, 12,187, 1987.

Sánchez, E.R., J.P. Thayer, J.D. Kelly, and R.A. Doe, Energy transfer between the ionosphere and magnetosphere during the January 1997 CME event, *Geophys. Res. Lett.*, *25*, 2597, 1998.

Solanki, S. K., M. Schüssler, and M. Fligge, Secular variation of the Sun's magnetic flux, *Astron. Astrophys.*, *383*, 706, 2002.

Subramanian, P., and K. P. Dere, Source Regions of Coronal Mass Ejections, *Astrophys. J.*, *561*, 372, 2001.

Thayer, J.P., and J.F. Vickrey, On the contribution of the thermospheric neutral wind to high-latitude energetics, *Geophys. Res. Lett.*, *19*, 265, 1992.

Thayer, J.P., Height-resolved Joule heating rates in the high-latitude E region and the influence of neutral winds, *J. Geophys. Res.*, *103*, 471, 1998.

Tsunoda, R. T., High latitude F-region irregularities: A review and synthesis, *Rev. Geophys.*, *26*, 719, 1988.

Wang, W., T.L. Killeen, A.G. Burns, and R.G. Roble, A high-resolution, three-dimensional, time-dependentm nested grid model of the coupled thermosphere-ionosphere, *J. Atmos. Terr. Phys.*, *61*, 385, 1999.

Wang, W., T.L. Killeen, A.G. Burns, and B.W. Reinisch, A real-time model-observation comparision of f_2 peak electron densities during the Upper Atmospheric Research Collaboratory campaign of October 1997, *J.Geophys. Res.*, *106*, 21,077, 2001.

Weiss, L.A., P.H. Reiff, J.J. Moses, B.D. Moore, and R. A. Heelis, Energy dissipation in substorms, *Eur. Space Agency Spec. Publ.*, pages ESA–SP–335, 309, 1992.

Chapter 9

Interdisciplinary Research

S. Adlen

9.1 Introduction

The mandate of the IUGG is the international coordination of studies of the Earth, its environment in space and analogous problems associated with other planets. Whilst the IUGG is nominally composed of seven, semi-autonomous bodies, by nature of the fact that the separate Associations study subsystems of an over-riding planetary system, interdisciplinary research is essential.

The IUGG supports interdisciplinary research in two ways. Firstly, through the establishment of joint working groups and secondly, through the formation of commissions.

Two or more International Associations may set up and support joint working groups to address specific scientific topics. During the lifetime of the IUGG, several important interdisciplinary collaborations have already been overseen with significant success, e.g.,

1. **The International Geophysical Year (1957-58)**

2. **The Upper Mantle Project (1964-70)**

3. **The Geodynamics Project (1972-79)**

4. **The Global Atmospheric Research Programme (1970-80)**

5. **The International Lithosphere Programme (1981-90)**

At the current time, there are several working groups in existence looking to further progress in geodesy and geophysics.

Another major achievement has been the creation, through the International Council of Science (ICS), of the World Data Center System.

Commissions are the second way in which the IUGG promotes interdisciplinary research. These are established by the IUGG to serve the Union and the international geophysical community by promoting the study of particular interdisciplinary problems and tend to be of a more permanent nature. These bodies are chartered by the IUGG Executive Committee to conduct inter-association science. Three Inter-Association Commissions are active at the moment:

1. **Committe for Mathematical Geophysics (CMG)**

2. **Study of the Earth's Deep Interior (SEDI)**

3. **Geophysical Risk and Sustainability (GeoRisk)**

The world today is markedly different from that of fifty years ago and is continually changing. A flexible framework in which interdisciplinary collaborations can be developed and disbanded as required, is needed. The real strength of the IUGG lies in its constituent associations, which together cover the full spectrum of international geodesy and geophysics and hence provide the requisite framework.

Within the remit of the IUGG, there are currently many opportunities for interdisciplinary collaborations. These opportunities fall into two main categories. There are both over arching opportunities in which all member Associations might contribute and more specialised fields which will be of interest to smaller groupings of Associations.

This chapter seeks to outline some of the major opportunities for inter-disciplinary research in the fields of geodesy and geophysics and to highlight potential roles for the IUGG in the realisation of these collaborative ventures.

9.2 Over arching Opportunities for Inter-disciplinary Research

9.2.1 Planetary Exploration

Comparative planetology and planetary exploration present the ultimate challenge as regards a problem that incorporates the full spectrum of geoscientific disciplines. The benefits of undertaking the challenges that are posed in pursuing these areas of research are largely two fold.

1. Comparative Planetology and the Protection of our Planet: The study of planetary geologies, atmospheres and other characteristics is important in our understanding of the evolution of the Earth and the basis of life on

Earth. By studying planets and other large bodies in the solar system, we learn about the history and possible future of our own. In exploring our solar system, we widen our perspectives and gain a better understanding of the mechanics of worlds in general, making us better able to protect our planet's fragile biosphere. Comparative planetology allows us to use other planets as full scale laboratories to improve our theories. As we move to a time in which Earth system modeling is becoming increasingly important, having the opportunity to test theories in other full-scale laboratories will be invaluable.

2. Widening the intellectual, physical and technological boundaries of Mankind: Exploration of other planets will push the technological capabilities of the human race and in doing so, will also expand our physical boundaries and understanding of the universe.

The Role of the IUGG

The IUGG is in a strong position to promote planetary research as it represents all geoscientific disciplines. The IUGG might consider the creation of a new association. An association which could better coordinate planetary exploration and modeling would be invaluable.

9.2.2 Mathematical Techniques

The geosciences provide an impressive array of important problems that command the attention of applied mathematics. They also offer a need, on top of an intellectual desire, to advance the foundations and techniques of applied mathematics to meet their challenges.

Methods in continuous and discrete dynamical systems, stochastic processes, homogenization, ergodic theory, renormalization, inverse problems and the quantification of uncertainty are all required. Some recent developments that require numerical and theoretical modeling tools are large-scale compressible jets (IAVCEI, IAGA), time series analysis and stochastic modeling of catastrophic events (IAVCEI) and tools, such as adaptive gridding techniques, to allow the incorporation of processes on disparate spatial and temporal scales into increasingly comprehensive models (IAMAS).

The Role of the IUGG

The Committee on Mathematical Geophysics is an inter-association committee of the IUGG, which already satisfies the need for co-ordination of research in the mathematical fields associated with the geosciences. The committee organizes bi-yearly conferences, and its continued presence is invaluable to the whole of the geosciences.

9.2.3 Observing Systems

All of the geoscientific disciplines require observations as a building block for their research. Combining these observations in the most thorough and efficient way is essential. Questions of both spatial and temporal resolution need to be resolved as each of the geoscientific disciplines seeks to improve their models to smaller and smaller scales.

Recently the problem of coordinated, integrated global observing systems has been addressed by the United Nations. Three inter-related global systems to observe the environment of the planet are being organized by United Nations organisations in co-operation with the scientific community and national governments.

1. Global Climate Observing System (GCOS): GCOS is responsible for planning the collection of data on long-term climate change.

2. Global Ocean Observing System (GOOS): GOOS is implementing operational observation programmes for the oceans and coastal areas.

3. Global Terrestrial Observing System (GTOS): GTOS is developing and networking observations of long-term change of the land and its resources.

An Integrated Global Observing Strategy (IGOS) is also being developed to improve coherence and joint planning for global observations of the environment. Whilst the above systems represent significant progress, observing systems that monitor the whole Earth system also need to account for the requirements of other geoscientific disciplines. Some examples of this are given below.

1. The Thermosphere requires improved monitoring.

2. Hydrologists require observations at a much higher spatial resolution than scientists concerned with climate change or the environment.

3. The whole of the Earth system is underpinned by accurate physical geodesy, which requires observations at increasingly higher spatial and temporal resolutions.

All desired observation goals and data products could easily be attained with a well-constructed global observing system. Some progress is already being made in using data from specific observing systems across a range of geoscientific disciplines (section 9.3.2). Both NASA's Earth Science Enterprise and ESA's Living Planet program are also significant steps toward reaching this goal *NASA* [1999]; *ESA* [1998].

The Role of the IUGG

The ICS is involved with the three observing projects described above hence already representing the IUGG to some extent. The IUGG should, however, ensure that the ICS is aware of its requirements as regards the data products returned from a global observing system that might benefit disciplines across the whole range of the geosciences. Whilst much political emphasis is placed upon the monitoring of the climate and environment, the needs of other scientific disciplines should not be neglected at its expense. Once a truly global observing system is in place then returning all of the data products needed for all realms of geoscientific research will be easily realisable. The IUGG should be proactive in the sponsorship of this goal.

9.2.4 Communication and the Availability of Data

The past decade has seen an explosion in information technology and a revolution in global communications. A major achievement has been the creation, through the ICS, of the World Data Center System, from which the data gathered during several major programmes are available to research workers everywhere.

The above type of undertaking is vital. All research requires data and the sharing and availability of data is thus essential. In the modern age of global communications, this should be relatively easy to achieve and would result in huge benefits to the scientific community.

The Role of the IUGG

The IUGG should advocate, and be active in promoting, free access to all data sets. The IUGG is also in a strong position to ensure that data is additionally made readily available to researchers in developing countries.

9.2.5 High Resolution Earth System Modeling

The ultimate challenge of Earth system science is to consolidate the scientific findings from different disciplines into an integrated representation of the coupled atmosphere, ocean, ice, land and biosphere system. It has become clear that models need to progress to higher resolutions in order to adequately represent processes taking place.

The Global Analysis, Integration and Modeling Task Force (GAIM) is a component of the International Geosphere Biosphere Program (IGBP) of the ICS. The goal of GAIM is to advance the study of the coupled dynamics of the Earth system, using as tools both data and models. This is just one example of a number of such projects throughout the world.

A further development in Earth system modeling comes from projects such as "The Earth Simulator Project" in Japan. This project creates a "virtual earth" on a supercomputer to show what the world will look like in the future by means of advanced numerical simulation technology. The reconciliation of processes occurring on vastly different spatial and temporal scales into a comprehensive Earth system model is non-trivial and demands input and co-operation across the whole range of the geosciences. This a growing field which will produce invaluable results to society over the coming years.

The Role of the IUGG

Each of the geoscientific disciplines feeds directly into an integrated Earth system model. The IUGG is in a strong position to coordinate Earth system modeling projects. Whilst the IGBP has already been established by the ICS, the IUGG could still promote the coupling of models for elements of the Earth system within the realm of the IUGG. Linking models of subsystems together one by one is an important process. The difficulties involved in trying to reconcile models that are based upon different spatial and temporal scales are enormous. If all of the subsystems represented within the IUGG could be coupled together then this would represent enormous progress and is something that the IUGG is in a position to actively encourage.

9.2.6 Looking to the Future

It is vital that the geosciences maintain a view a of future with the continual re-assessment of goals and milestones. Undertaking projects that push our intellectual and engineering capabilities to their limits are important in the advancement of research. Projects such as the development and maintenance of new habitable environments and the intentional forcing of systems to mitigate hazards could be established to the benefit of both society and research. Through this type of undertaking enormous progress is often made in more routine areas of research. Aim for the stars, reach the sky!

The Role of the IUGG

The IUGG general assembly every four years provides an excellent platform for reflection and re-assessment. The sponsorship of working groups such as "Geosciences: The Future" and attaining ideas of young scientists is an excellent idea. Through the promotion and co-ordination of forward thinking and far-reaching research ideas, the IUGG can be of huge value to the geoscience community.

9.3 Other Opportunities for Interdisciplinary Collaborations

In addition to the opportunities for interdisciplinary research which encompass all fields of geoscientific research, there are many chances for research between a smaller number of the associations of IUGG. Some of the potential opportunities, which are currently being, or could be exploited, are presented below.

9.3.1 The Hydrologic Cycle (IAMAS, IAPSO and IAHS)

The International Association of the Physical Sciences of the Oceans (IAPSO), the International Association of Hydrologic Sciences (IAHS) and the International Association of Meteorology and Atmospheric Sciences (IAMAS) have the chance to work collaboratively to understand the dynamics and interaction between ocean, atmosphere, surface and groundwater, which are closely linked via the hydrologic cycle. This is a difficult task that involves the reconciliation of models on vastly different spatial and temporal scales. The task is, however, crucial to the maintenance of water supplies and hence life on Earth.

During the IUGG general assembly in Sapporo, several interdisciplinary undertakings are already being investigated related to the hydrologic cycle.

1. **The Decadal to Centennial Variability of the Ocean and Atmosphere (IAPSO, IAMAS, CLIVAR)**

2. **The Surface Ocean-Lower Atmosphere Study (IAMAS, [ICACGP,ICPM], IAPSO)**

3. **New Perspectives of Coupled Tropical Ocean-Atmosphere Dynamics and Predictability (IAPSO, IAMAS, CLIVAR)**

4. **Arctic Environment Change (IAMAS, IAPSO, IAHS)**

5. **Groundwater Inputs to the Ocean (IAPSO, IAHS)**

6. **Snow Processes: Representation in Atmospheric and Hydrological Models (IAHS/ICSI, IAMAS)**

7. **Water and Energy Budget Workshop (IAMAS,IAHS)**

8. **Measurement and Distribution of Precipitation (IAMAS, [IPCC, ICPM], IAHS)**

9. **Worldwide Sea Level Change (IAPSO, IAG)**

10. **The Treatment of Precipitation in Cloud and Climate Models (IAMAS, [ICCP, ICCI],IAHS)**

11. **Land Ocean Atmosphere Interaction in the Coastal Zone (IAMAS, IAPSO, IAHS)**

12. **Global Sea Level Rise, Global Climate Change and Polar Ice Sheet Stability (IAMAS, [ICDM, ICCI, ICPM], IAPSO, IAG, IAHS)**

13. **Physical Aspects of Air-Sea Interaction (IAPSO, IAMAS)**

Improving and coordinating these investigations so as to be more strongly linked with the over-riding goal of understanding the hydrologic cycle would be beneficial. There is scope for the IUGG to endorse these studies through the development of a structured programme of research.

9.3.2 The Role of Geodetic Sensing in Interdisciplinary Earth Science (IAG and Others)

Classically, mapping has been viewed as the main purpose of geodesy. However, by establishing reference geodetic networks, geodesy has become fundamental in the performance of other disciplines, whose activities are related to positioning, such as urban and environmental management, engineering projects, (international and national) boundary demarcation, ecology, geography, hydrography, etc. Many applications are also found in other scientific fields, e.g., theoretical and exploratory geophysics, geodynamics, space science, astronomy, oceanography, atmospheric science and geology. New techniques to observe and to analyse the Earth as a complete system have made the relationships between disciplines stronger. Accordingly, some interdisciplinary enterprises that utilise or could utilise geodetic information are highlighted below.

1. **The Hydrological Cycle (IAG, IAMAS, IAPSO, IAHS, IASPEI)** Geodesy can provide precise information about melting (glaciers and polar icecaps), sea level change, atmospheric mass movements, ground water and soil moisture.

2. **Precise modeling of crustal deformation, including plate tectonics, volcanoes and earthquakes (IAG, IAVCEI, IASPEI)** Geodesy can precisely monitor crustal deformation (secular and episodic, extended and local), as well as gravity and Earth's rotation changes due to mass displacements. The deformation of surface topography reflects the dynamical processes that are occurring in the Earth. Continuous geodetic observations of surface deformation provide not only useful information for understanding the dynamics of the Earth's interior, but also indispensable information on the occurrence of earthquakes and volcanic eruptions. Most volcanic activities are accompanied by seismic activity. Revealing processes of interaction between earthquakes and volcanic activities would be a clue to better understanding the triggering effects between eruptions and earthquakes. To obtain detailed images of the seismic structure of volcanoes, enhanced seismic observations in volcanic areas as well as the advancement of theories and techniques of seismic wave propagation and modeling are essential.

3. **Atmospheric Modeling (IAG, IAGA, IAMAS)** Geodesy can provide valuable information about pressure and humidity from atmospheric sounding using GPS.

4. **Coupling Between Atmosphere, Ocean and Solid Earth (IAMAS, IAPSO, IAG)** How do the solid Earth, atmosphere and ocean interact with each other? How do the atmosphere and ocean shake the solid Earth? Does the melting of polar ice caps due to global warming induce the deformation of the surface topography?

   The development of a systematic Earth model integrating a variety of Earth processes, such as global warming, growth and regression of polar ice caps, plate tectonics, continental evolution, etc, which fully incorporate the physics and chemistry of the processes considering the significant differences in time-scale, is a common goal of geosciences. This is a piece in the overall goal of a complete Earth system model (section 9.2.5).

It is important that as much information as possible is extracted from geodetic observations. Whilst this highlights the fact that not enough information is currently extracted from available observations, it is, however, just a small issue in the ultimate goal of achieving a global observing system (section 9.2.3). Sessions on "Interdisciplinary Science from Improved Earth System Modeling" and "Interdisciplinary Science from Remote Sensing" in Sapporo are positive steps toward addressing this problem. The IUGG will continue to play an important role in addressing both the issue of efficient data usage and the implementation of a truly global observing system.

9.3.3 The Interaction of the Thermosphere, Upper Atmosphere and Solar Forcing (IAGA and IAMAS)

The link between phenomena in the Upper Atmosphere and Thermosphere and the additional role that solar forcing plays in both is an active area of research between the associations of IAGA and IAMAS.

The mesosphere-thermosphere-ionosphere (MTI) system is affected by long-term changes and trends in such things as atmospheric concentration of greenhouse gases, geomagnetic activity and solar activity. The phenomenology of the equatorial ionosphere-thermosphere system is controlled particularly strongly by vertical coupling to lower atmospheric layers. Understanding the coupling between these systems is essential. An additional connection between the associations of IAGA and IAMAS is the fact that climate change could partially be of solar origin.

The relation between the Upper Atmosphere, Thermosphere and solar forcing is an important aspect that must be included in any full Earth System Model. The IUGG is already aiding this area of research through several joint symposia at the general assembly and continued support from the IUGG is required. The working group on Long-term Trends in the Mesosphere, Thermosphere and Ionosphere already goes some way to supporting this field of research.

9.3.4 Climate Response to SO_2 Aerosol Forcing (IAVCEI, IAMAS)

The emission of SO_2 from volcanic eruptions provides an excellent opportunity for atmospheric scientists to study the response of the atmosphere to a point-like forcing and for vulcanologists to investigate the processes through which these volatiles are released into the atmosphere. The study of these events requires significant collaboration, particularly in terms of chemical modeling, prediction and observing systems. A joint symposium has been arranged by the IUGG to bring together expertise from both fields with aim of tackling this problem. This initial step should be built upon to the long-term benefit of both associations.

9.3.5 Generation and Propagation of Tsunami (IASPEI, IAPSO)

Earthquakes in oceanic regions sometimes generate huge tidal waves, "tsunami," which bring catastrophic damage in coastal areas. The scientific investigation of their generation and propagation is central to the creation of a practical warning system to mitigate tsunami disasters. The commission on Tsunami is already active in advancing research in this field.

9.3.6 Seismology for the Earth and the Sun (IASPEI, IAGA)

The studies of free oscillations of the Earth and the Sun shares some common theoretical grounds. The development of seismology would complement helioseismology, and vice versa.

9.3.7 Water, Earthquakes and Volcanoes (IAHS, IASPEI, IAVCEI)

The process of groundwater transportation and its implications to the precursory anomalies of earthquakes need to be evaluated quantitatively. Fluids in the crust may play a crucial role in earthquake and volcanic dynamics.

It has long been known that fluid injection can generate seismicity. In addition, within the last decade, it has been discovered that seismic waves can trigger earthquakes in geothermal areas and many explanations involving mobile, hot fluids have been proposed. Fluids are also often invoked to explain the low friction on faults and other dynamical features of fault rupture.

Now that solid Earth geophysicists have posed these questions about the role of crustal fluids, they need to answer the problems using the knowledge-base of the hydrological sciences. Ground-water monitoring could constrain the mechanisms for long-range triggering and changes in water chemistry and temperature may constrain the degassing history of shallow magma bodies. The dynamic role of water during rupture can, however, only be quantified if we also understand the groundwater flow in the vicinity of the fault both during and in between earthquakes. Many of these questions will also push the boundaries of hydrology.

9.3.8 Interactions with Disciplines Outside Geoscience

In addition to the potential for collaborations within the geosciences, there are also many opportunities for interactions with disciplines from outside the realm of the geosciences.

1. **Social and Political Sciences** In association with the social and political sciences, questions such as how the geosciences might best serve society should receive attention. For example, issues of early warning systems, sustainable development and the support of science must dealt with.

2. **Biological Sciences** As Earth system modeling becomes increasingly important, so the understanding of the interaction of biological systems with the geo-environment becomes essential. Quantitative understanding of biological exchange processes, for example in the carbon cycle, is a significant unknown in Earth system models.

3. **Engineering Sciences** The relationship between engineering science and the geosciences occurs on two levels. From a positive viewpoint, engineering science provides the capacity for improved resilience to geo-forces. On the other hand, engineering projects can have a detrimental effect on geo-systems, e.g., pumping that causes subsidence. The relationship between the engineering and geosciences should be strengthened in order to ensure that our engineering capabilities are used most beneficially for society.

9.4 Conclusions and Recommendations

There exist many exciting opportunities for interdisciplinary collaborations within the geosciences. The IUGG is already assisting in the promotion of several of these opportunities with other worthwhile cases available which would benefit from the support of the IUGG in the future.

Bibliography

ESA, , The science and research elements of ESA's living planet programme, Technical report, ESA, 1998.

NASA, , EOS reference handbook, Technical report, NASA, 1999.

Oki, T., The global water cycle, In Browning, K., and R. Gurney, editors, *Global Energy and Water Cycles*. Cambridge University Press, 1999, pp.10-27.

Chapter 10

Societal Impact

K. Yoshizawa

10.1 Introduction

The Earth provides us with a number of benefits, such as natural resources and an inhabitable climate, which help to enrich human life. However, it sometimes can harm us with natural disasters and severe weather. Sustainability of human society depends strongly on the environment, and thus the development of geosciences is critical for a better understanding of our living environment, and its sustainable development.

Scientific results, particularly those related to natural hazards and the human environment, should be transferred properly to society, decision and policy makers, and the public, addressing issues and providing guidance in response to specific societal needs.

Action-oriented recommendations from the geoscientific community can be used to evaluate, formulate and create a plan of both national and international policies for sustainable development of our society. Such an effort will be crucial for the communities of geosciences to strengthen their credibility, which will be helpful in enhancing the possibilities of gaining better financial support for further development of geosciences. It is also an important role of international scientific communities to make efforts to reinforce the capacity of developing countries to deal with environmental, social and economic issues.

IUGG stands in an important position for promoting scientific studies of the Earth and its environment and for communicating with the international society. Therefore, the IUGG can be an ideal facilitator to link the people's interests and state-of-the-art geoscientific researches. In this chapter, we discuss some important aspects of the relationship between society and geosciences, and what geoscientists can do for the benefit of human beings living on this planet.

10.2 Geophysical Risks: Prevention of Natural Hazards

The uniqueness of the Earth compared to other terrestrial planets is its dynamic activity in a variety of forms, i.e., global circulation of the atmosphere and oceans, plate movement, volcanic activities and earthquakes, which sometimes cause devastating damage to our society. Society is affected both economically and socially by the impact of such natural hazards. Even if our life standard has been improved by virtue of technological development, there seems no practical way to keep away from such disasters. Although we are always going to be at risk from geohazards, it is certainly possible to limit the damage they cause. Anything that can be done in this respect is very worthwhile.

10.2.1 Interactions Between Community and Geosciences

Close Linkage with Decision Makers: Scientific progress will not mitigate hazards unless results are transferred effectively to local decision makers during a crisis. Usual practices in communication are inevitable for both scientists and decision makers to take prompt actions against geohazards. In order to make the most of the latest results of geosciences, it is essential to establish a means of interactive communications between decision makers, public and geoscientists. All important information must be passed to the public in a common language that can be understood by anybody. Only with such efforts can the latest results of geosciences be a useful guide for the public. Through communication with policy makers, society becomes aware of the potential risks and can account for them in terms of adequate insurance coverage and defense against the potential hazards.

Efficient Use of Hazard Maps: To evaluate potential risks of geohazards, hazard maps based on geological and geophysical researches are of great help for employing countermeasures against natural disasters. Also, it can be useful for creating practical manuals in case of an emergency, for the establishment of insurance policies, and for future land use. This type of information is particularly useful for avoiding devastating disasters that are still difficult to predict precisely, such as volcanic eruptions, earthquakes and tsunami.

Employing Early Warning Systems: With extensive observation systems, it is now possible to transmit an emergent warning to the public immediately after a natural hazard has occurred. For example, in the case of large earthquakes, an automatic shutoff system for gas lines is important to prevent subsequent fires. Particularly for mega-cities with growing populations, establishment of an early warning system is essential to mitigate the damages of geohazards.

10.2.2 Effective Use of Technology and Scientific Results

Comprehensive Observation for Hazard Mitigation:
Continuous monitoring of the Earth is one of the most essential factors to mitigate natural disasters. Observation with satellites provides us with many useful information about the current state of the Earth at both global and regional scales, for example, monitoring rain falls, sea level rise, variation of gravity fields and so on. GPS observation provides us with information on the crustal deformation, which may lead to fault rupture or volcanic eruptions. Continuous geodetic observations of crustal deformation may enable us to locate possible earthquake sources. The measurements of the gravity fields can be helpful in detecting mass movements in the Earth, e.g., movement of a magma chamber beneath volcanic areas, which may lead to an immediate eruption.

Information Technology: In case of devastating natural disasters, such as earthquakes and tsunamis, efficient and stable ways to transmit the situation in a disaster area are indispensable for taking prompt actions against the hazards. It will be useful to create a system utilising information technology as a communication tool in case of emergency. Outer space monitoring for disasters and damages may be helpful in the immediate transmission of rescue information.

Improve Predictability: The current situation of the predictability of natural hazards depends significantly on the type of natural phenomena, e.g., the occurrence of severe weather can be well observed and predictable through comprehensive means of monitoring the state of the atmosphere and hydrosphere, whereas disasters rooted in the geosphere, such as volcanic eruptions and earthquakes, are rather difficult to predict, since the direct monitoring of the Earth's interior is impossible. By improving models of the system with which the risk is associated, and enhancing monitoring system, predictability can be improved, thus allowing society to be better prepared when the hazard strikes.

Simulating Geohazards: It will be helpful to utilise new powerful computing facilities, which enable us to simulate the effects of natural hazards with high degree of accuracy. Especially, for some gigantic disasters with difficulty in their predictions, such as earthquakes, this approach should be helpful to quantitatively assess possible damages. The results of numerical simulations can be a useful guide for taking precautions and countermeasures against devastating natural disasters.

10.2.3 Some Counter-measures Against Gigantic Natural Hazards

Societal impact of some typical natural disasters are summarised below, along with raising some plausible countermeasures against these hazards.

Extreme Weather Events: Extreme weather events are some of the planet's most devastating natural disasters. There are many forms of meteorological phenomena at a variety of scales of both time and space, which sometimes bring devastating damage to society, e.g., typhoon, hurricane, tornado, downfall, flood, drought and so on. Progress in the study of the atmospheric interactions that shape weather phenomena has created an opportunity to make major advances that have led directly to improved weather warnings and predictions. This gives communities a greater confidence, an improved chance of minimising the effects of extreme weather events, and hence a superior quality of life. New atmospheric models have the capability to run at horizontal resolutions of around 1km or less, with the increasing power of computational facilities, and open up the possibility of significantly improving predictions of severe weather. Elucidating relationships between extreme events and modes of climate variability, such as El Nino, will also be vital.

Volcanic Activity: Most of volcanic activities are accompanied by some precursors, which can be observed in most volcanoes as sudden increases of seismic activity and/or crustal deformation. These events are likely to be related to active movements of magma chambers beneath volcanoes, which can be associated with subsequent eruptions.

Prediction for volcanic eruptions seems more plausible compared to that for earthquakes. In fact, there have been successful examples for predicting eruptions, which saved a number of lives (e.g., an eruption of Mt.

Usu in Hokkaido, Japan in 2000). However, these successful predictions seem to be rather empirical, and the long-term experience of observers is required for making decisions about warning society prior to the eruptions. More quantitative methods for evaluating the state of magma chambers underground will be inevitable. Continuous and massive monitoring of volcanoes with high potential risks of eruption should be essential.

Earthquakes: Preparing for catastrophic earthquakes is a critical issue for large cities with high seismic activity. Practical ways to predict earthquakes are not yet available. However, it is invaluable to take precautions against disastrous damages that may be caused by large earthquakes.

In order to prevent hazardous disasters in urban areas, especially such as mega-cities with high population densities, it is vital to quantitatively evaluate potential seismic risks, utilising hazard maps based on geological and geophysical investigations of active faults and local underground structure in conjunction with numerical simulations of strong ground motion. Basic researches on crustal deformations and seismicity pattern analysis are also essential.

From the view point of societal benefit, the legislation of appropriate building codes is vital in areas with high potential risks. It is also important to utilise real-time information on strong ground motions for controlling large structures in order to minimize the impact of large earthquakes on metropolitan areas.

Tsunami: Some earthquake sources in the ocean bottom and land slides near coastal areas occasionally cause large tidal waves, tsunamis, which hit coastal inhabited areas, bringing catastrophic damages to cities facing the oceans. In case of shallow earthquakes in the ocean bottom, the generation of tsunamis can be expected. Thus, the immediate determination of an earthquake source helps for early warning of tsunamis, which gives communities great confidence, minimising the damages of tsunamis. With detailed information on bathymetry, it is possible to simulate propagation of tsunami waves, which helps to create tsunami hazard maps. Construction of breakwater based on this information is essential for preventing hazardous tsunami disasters.

10.2.4 The Role of the IUGG

In 2000, IUGG launched the Commission on Geophysical Risk and Sustainability (GeoRisk Commission) for promoting scientific studies to reduce the risks of natural hazards in the modern society. The continuation and growth of this commission allows the IUGG to play a significant role in the management of georisk.

10.3 Human Life and Environment

For improving the quality of life for human beings, a better understanding of the Earth system is invaluable. Continued studies of geosciences will be essential to keep the continuous and sustainable development of modern society and to keep a balance between human welfare and the natural environment.

10.3.1 Global Change and Human Activity

Global warming is one of the critical issues for modern society. Continuous temperature rise of the atmosphere will cause melting of polar ice caps, resulting in the sea-level rise, which will threaten most metropolitan cities near coastlines all over the world. For some countries in the southern Pacific islands, the effect of the sea-level rise is a vital issue, which threatens the existence of their home lands. The cause of the average temperature rise can be attributed to human activities, especially the rapid industrial developments throughout the 20th century. Still, it is important to separate the effects of human activity from natural global changes, so that we can quantitatively evaluate the limitation of industrialisation. As certainty of the causes for climate change improves, the thresholds of tolerable risks due to projected climate change need to be defined, and the communication of scientific results to both society and policy makers needs to be improved.

Anthropogenic Effects: Global climate change has attained the enormous interest of the public due to its potential to effect human activities and the environment. A critical issue is to understand the possible anthropogenic role in climate change. Human-induced changes in climate, observed on decadal time scales, are already evident. Atmospheric pollution now hangs over many regions of the Northern Hemisphere. The thinning of the stratospheric ozone layer, most notably over Antarctica, is a human-caused feature of the planet. The greenhouse gases being added to the atmosphere will reside there for decades to centuries and are predicted to increase average global surface temperatures by several degrees, a change that is larger than the natural variation occurring over the past 15,000 years. The separation of natural climate change from that caused by the activities of humans needs to be established.

Observation System: Satellite altimetry enables scientists to forecast El Nino, whose influence on the global climate as well as on agriculture, agronomy and hydrology, is enormous. The sea level changes, measured also with satellite altimetry, are a response to global warming. The influence of the global climate changes on ice sheets can also be observed in a similar way. All of these aspects form an accurate and reliable tool for assessing the current climate and for predicting future climate changes in both short and long terms as well as appropriate warning to countries and cities facing coastal areas.

Modeling Climate: Only through the development and application of a high-resolution integrated global modeling and observation system can our influence on the climate and environment be estimated. A better understanding of sources, processes and couplings between different parts of the climate system is required so that anthropogenic effects can be accurately and efficiently represented in models.

Role of Geoscience in Environmental Issues: The geosciences, especially atmospheric sciences, oceanography and hydrology, are a vital component in the control of environmental issues, e.g., in the monitoring of air and water pollution and greenhouse gases. It is important to transmit the latest information to warn the public about the influences of human activities on environmental problems.

Increased Public Awareness of Issues Affecting Our Climate: One of the most important factors of geosciences is the communication to the general public of the effects that humans are imposing upon our climate. As certainty in causes of climate change improves, these results must be communicated, as a matter of urgency, to policy makers.

Political Aspects: Global environmental issues can partly be attributable to the for-profit disposition of the world's societies. It is important to keep a balance between the preservation of the natural environment and the continuous and sustainable development of society so that we can sustain growing populations, especially in the developing countries. Precise understanding of the effects of human influence on the environment and the communication of these results to policy makers are essential.

10.3.2 Management of Water

Water is one of the most fundamental resources necessary for all life inhabiting the Earth. In some parts of the world, the constant shortage of water resources is a vital issue, which directly threatens human life. The management of water resources is an environmental issue, which results from geoscientific researches. Atmospheric physics, hydrology and oceanography, have an integral role to play.

Hydorological Cycle in the Earth: It is essential to strengthen international cooperation for the study of processes related to the Earth's atmosphere, hydrosphere, biosphere and geosphere as a way to better understand water resources and the movement of water through the Earth system.

Infrastructure for Stable Water Supply: The infrastructures for water supply is not yet sufficient in many areas, especially in developing countries. In order to sustain the growing population in the world, international cooperation and support for a stable water supply and sanitation are essential to provide fresh and clean water, particularly in developing countries. Prediction and management of water resources is of vital importance for the sustainable development of society.

Transferring Scientific Achievements: Only if the scientific communities work together with policy and decision-makers can water issues be successfully addressed. Knowledge and understanding of water issues are essential and should be further encouraged. National and international information networks on water issues, using modern technologies, must be strengthened. This is especially crucial in many developing countries where international assistance needs to be enhanced. It may be helpful to formulate communication strategies that would allow scientists to better communicate with society disseminating their latest results. It should be vital to respond to the needs of policymakers while being firmly based on scientific understanding and knowledge.

10.3.3 Improvement of Quality of Life

The quality of human life can be improved by the effective use of scientific results on the Earth system. Several geoscientific topics that enrich our everyday life are discussed below.

Weather Forecasting: Reliable forecasting is becoming increasingly important to society, especially regqarding its effect upon energy, health and agriculture. Accurate weather forecasting allows a greater certainty when planning ahead in weather dependent industries. It also benefits the management of water resources and other elements having a direct influence upon the sustainability of both our environment and our economies. Skill in forecasting has improved dramatically over the last 20 years, and today, we enjoy considerable confidence in predictions from periods of days to years.

Near-Earth Space Environment: Satellites, which are essential instruments for observing the Earth, are now also indispensable to our everyday life as communication tools on the Earth. However, they cost a lot of money to build. Satellites can be destroyed by a number of phenomena, which occur in the near- Earth space environment, such as radiation erosion of electronics, charging of space craft and subsequent arching, which may occur, and the ejection of the space craft from the magnetic stability region of space. Understanding of the near-Earth space environment is quite important because of the increasing number of space assets, which incur significant costs in their building and launching.

Precise and Reliable Reference Frame: The combination of the physical and geometrical reference systems, together with the improvement of their components, will support the users to achieve 1 cm level in navigation and to increase up to a few centimeters level of the real time positioning accuracy. This will also help to build a consistent infrastructure of global spatial data under the new concept of the "digital Earth."

Energy Resources and Sustainability: We have been relying mainly on fossil fuels, such as oil, coal, and natural gas, as the major sources for creating energy, which is consumed enormous amount everyday. Although fossil fuels have been sustaining modern society, their use has yielded drastic changes of the Earth's climate due to the emission of green house gases, which leads to global warming. Besides, there is a limitation in the storage of such fossil resources in the Earth. For continuous development of the human society and, as a whole, preserving the natural environment, the quest for alternative energy sources is essential. As a possibility, wave energy of the ocean can be a clean and effective energy source. The recommendations for the use of other clean sources, such as solar and wind power, should also be promoted worldwide.

10.3.4 The Role of IUGG

IUGG is in the most appropriate position to be a facilitator to try to adjust the balance between development and environment, making the most of its truly international character and expertise in geosciences. Proactive actions of IUGG for communicating with society will be more important for sustainable development of international society in the future.

10.4 Geoscience Education and Outreach

Geosciences are among the most important scientific studies, which are directly related to our living environment and have a significant impact on the modern society. For the continuous development of geosciences in the future, promoting education at various levels is essential. In recent years, children's incuriosity about science has been a serious concern of modern society. It is essential to enlighten people about the current state of the Earth, and what we can speculate about the future of our environment from the latest results of geosciences.

10.4.1 Efficient Use of Modern Technology for Scientific Education

IUGG and its associations have been publishing educational materials for school use. Still, most developing countries are suffering from shortages of qualified teachers in geosciences. Recent rapid growth of internet technology provides us with the possibility of a world-wide education system through the internet.

In this regard, enhancing the contents of the IUGG web site is fundamentally important to promote the scientific activities of IUGG, as well as to attract people's interest in geosciences. It is also recommended to translate the contents into some major languages other than English and French. It may also be helpful to upload some educational materials on the IUGG web site, so that people in any country can easily access them.

10.4.2 Promoting International Schools

IUGG should also facilitate international schools of geosciences for young generations. It should be useful to open IUGG summer schools, for example, for K-12 students, undergraduate and graduate students, and for the public, inviting lecturers from each association to foster the geosciences. In this regard, establishing an international education centre for geosciences may be of great help in fostering geosciences to the next generation.

10.4.3 Communication with Society

A critical issue for the IUGG to consider in the coming decade may be to look for better ways to convert and disseminate scientific information for societal use. One area of improvement could be a better education of the public through more convincing arguments about the importance and usefulness of science, especially in dealing with the assessment and mitigation of natural hazards. Television is a great tool in this respect; therefore, more educational television programs should be developed to address such issues. Communications with mass media at regular intervals could also be useful for dissemination and preservation of scientific results.

10.4.4 International Assistance to Developing Countries

International cooperation is inevitable for bringing up young geoscientists and promoting geoscience education and researches in developing countries. In this regard, professional training programs for scientists from developing countries are useful to better understand issues they face in their countries. Appropriate scientific and technological assistance to developing countries is also vital in order to reduce damages and casualties in these regions in case of disasters.

10.5 Conclusion

The role of geoscientists will become more important in the 21st century to foster people's understanding of our planet. Because of its truly international character, IUGG has an advantage over other large geophysical associations, which are oriented toward a particular country or region. The role of IUGG as a facilitator to link the international society and scientific community should become more important in the coming decade for the sustainable development and the preservation of the natural environment.

10.6 Acknowledgment

The author would like to thank B. Chouet and K. Suyehiro for their useful comments on issues of IUGG for societal benefit.

Chapter 11

Developing Countries

L. Sánchez

Geosciences have strongly developed in the last century, not only in observation techniques, but also in theory and modelling. Their analyses are key in the quest towards a clearer understanding of the Earth, its natural resources and environment, and human interaction. However, this development has been negatively affected by its non-homogeneous appropriation world-wide. The phenomena that are studied by geosciences have a global character, they are present over the whole Earth, and their variations affect also the whole Earth, so they have to be uniformly observed over the whole Earth. In the same way, the advancements of the new discoveries in geosciences have to be globally applied to continue a harmonized development, always looking for the maximum benefit to society, seen also from the whole Earth point of view. Nowadays, it is not possible to assume geosciences as isolated islands, they are strongly connected and their specific analyses and complement each other. Similarly, the isolated study of geoscientific phenomena in space and time leads to misinterpretations that, in the end, will represent inaccuracies in the corresponding models and results. Until now, the primary role in the performance (development, application and benefits) of geosciences is being played by industrialized countries; however, the state-of-the-art demands that developing countries interact directly on knowing, developing and taking advantage of the new global character of geosciences. According to this, the present chapter will discuss the principal aspects of the relationship between developing countries and the geosciences covered by the International Union of Geodesy and Geophysics (IUGG).

11.1 Natural and energetic resources provision

The growing population (more than six billion people) and global economies place enormous pressures on natural resources and environments. As a matter of fact, the energetic and natural resources reservoirs are mainly located in devel-

oping countries, and its exploitation generates strong damages in the environmental conditions. This situation places geosciences in a strategic relationship with developing countries: Firstly, preservation and adequate exploration of these resources require new scientific knowledge and advances in technology to facilitate their management. Secondly, they allow enhanced environmental stewardship and make it possible to avoid or minimize potential catastrophic problems, such as famine, waste, and environmental degradation. Keeping in mind that these resources are fundamental to life world-wide, their consequences are also a world-wide responsibility. The challenges of this century, among others, include how to stabilise the fast-growing global population, sustain food production, avail all of sufficient fresh water, minimise the impacts of natural disasters and address environmental concerns. This requires new commitments in science and technology not only from industrialized countries, but also from developing ones. In this regard, the field of geosciences, relevant to the expertise of the IUGG, should contribute to addressing the concerns and challenges of society.

11.2 Observing data world-wide

Recent developments in science and technology, such as computers, communication and information technology, satellites and new numerical methods, have greatly contributed to the monitoring, collection, processing and distribution of geodetic and geophysical data and products for operational and research activities. In the same way, the Earth is observed as a unique system with an extended gamma of phenomena that are strongly correlated and obligate the use of input data in a global scale with the convergence of all geosciences. However, the observatories to these phenomena are mainly located in the industrialized countries, and its poor densification in developing countries (the greatest part of the globe) leads to inaccurate models that create obstacles for the continued advancement of geosciences. As a consequence, it is necessary to improve the local observing infrastructure (equipments, procedures, modelling) in developing countries. Keeping in mind that these countries do not have sufficient financial resources, educational features or expertise to undertake this task, IUGG should play an orientation role, promoting international projects with technology and educational transfer, making possible the integration of geoscientists from developing countries in special working or study groups, organizing scientific meetings in these countries to advertise geosciences among them, guaranteeing free and cheap publication facilities, making data available for scientists in developing countries.

11.3 Strategic positions to analyse concrete phenomena

The geophysical variables in developed countries have strongly been studied, and their new developments are related to improving already high accurate models (climate, gravity, magnetism, etc.) over these regions. However, the modern observing techniques, especially satellite measurements, have demonstrated that some models present poor quality in a global scale. The new observations show special and unique geophysical features, that are located in regions of developing countries. They define the geophysical behaviour of the whole Earth, but they have been neglected in many precedent models, because they were not observed. This leads to building a homogeneous in situ observation set that permits to control and improve the global observation networks and the space-borne measurements, making accurate global models reliable. From this point of view, the participation of developing countries in the improvement of geodetic and geophysical models has to be direct, not only by supplying field observations to scientists of industrialized countries, but also by specialising their knowledge on the particular conditions. Examples are El Niño, electro jet effect, strong gravity gradients, volcanoes and geological faults, etc.

11.4 Importance of geosciences in the modern society, and especially in developing countries

Although a lot of work has been done to promote geoscientific information to the public in many countries, population and administrators have very limited knowledge of geosciences, and of its importance inside society. In developing countries, the popularization of geosciences is worse than in the industrialized ones, being limited and hindered by many factors. Two important aspects are, firstly, the restricted purchasing power of people and the relatively small investments in education, culture and technology, and secondly, the decreasing public support for research funding in geosciences. The use-

ful tools produced by geodesy and geophysics to natural disasters management, planning, sustainable development, etc. are not known, and as a consequence, these disciplines are neglected. In this way, the question of public awareness of geosciences is a very relevant and global theme for all those who work with geodetic and geophysical research. At this point, the role of IUGG is fundamental in providing the public, decision makers and industry with correct information on themes related to geosciences and its professionals. However, these questions are not restricted only to developing countries, but they also hold for the industrialized ones.

11.5 Vulnerability to natural disasters

The natural and social conditions in developing countries make them highly vulnerable to natural disasters, such as earthquakes, flooding, volcanic eruptions, landslides, drought, severe weather events and changes in climate, which frequently affect the poorest parts of the population. One of the most important objectives of geosciences, as a consequence of IUGG, should be to supply technical tools that lead to mitigate these natural hazards. In this way, it is desirable to involve native scientists in large international initiatives, because they know the problem and its particular features. The benefits will be reciprocal: developing countries could implement actual effective policies to control the natural hazards, minimizing the number of victims, and geosciences could take maximum advantage of these natural tragic events to improve their theories and models.

11.6 Sustainable development

In the new century, the scientific disciplines have renewed their compromise to serve to improvement the quality of life for mankind. The industrialized countries concentrate their efforts in controlling climate and environmental problems to maintain their high standards of life; however, they are sharing the same environment with non-developed countries, whose alternatives to follow the same way are inexistent. Overlooking the current problems in developing countries will generate environmental damages on a global scale. In this way, one of the main purposes should be to enable people and organisations in these countries to improve their living conditions through their own efforts, alleviating poverty, protecting the environment and natural resources and improving educational quality. Geosciences, under the umbrella of IUGG, should promote the planning and implementation of projects and programmes of technical cooperation in the areas of its Associations, the main tasks being soils, water, environment, mineral resources, energy resources, mining/waste deposits as well as geohazards. The overall goal is the satisfaction of basic needs, sustainable economic development, management of natural resources and protection from geo-hazards, as well as sector related spatial and regional planning.

11.7 Transferring technology and improving scientific level

The best way to improve the performance (knowledge, development, applications) of geoscientific research in developing countries is through teaching, research and services, increasing new geoscience ideas and concepts and preparing scientists of these countries. Specialists are needed in linking scientific understanding of resources and environments with other disciplines, such as public policy and law, to ensure proper land use planning. However, this should be done in a critical manner, methods used in industrialized countries should not be applied directly to developing countries. They should be adapted to the specific conditions and characteristics of the region and its people. This should be done in strict cooperation with the local scientists and entities that are working in geosciences. In this way, IUGG should assist, in particular, the developing countries in the training of operational personnel and research scientists, and in the dissemination and use of research outputs. IUGG should also take a leading role in ensuring free and unrestricted international exchange of geophysical data and products, and ensure the availability of scientific journals and other research results related to geosciences and to the developing countries.

The solution of the problems in developing countries requires interventions from Earth sciences. IUGG should participate in this challenge and develop strategies for the reliable assessment of natural resources and for optimising their rational use. To support this process, progress in geosciences and modern observing technologies should be used to develop appropriate databases to support research and operational activities through international initiatives that invite the active participation of developing countries.

11.8 The role of developing countries in geosciences

The above items summarize some important aspects that IUGG should address to support the performance of geosciences in developing countries. However, this cannot be a "one-way" process; the developing countries have to understand how important geosciences are to improve their disadvantageous conditions and try to participate more actively in the new challenges. Unfortunately, geosciences are not very popular among policy makers, administrators and the public in general, not only in the industrialized nations, but also in the developing countries, where the applicability of geosciences and their products is almost totally unknown.

In these countries, normally, professionals working in geosciences do not know about the existence and advantages of international organisations like IUGG or its Associations, who are offering a world-wide setting to improve their daily activities. At this level, the scientists of developing countries, who are involved in the IUGG (or its Associations), are requested to promote among their national (regional) colleagues the offered opportunities to participate in the international structure of geosciences, improving directly their experience, knowledge and quality results. The traditional conviction of the industrialized countries about the insufficient capability of developing countries to assume important roles in scientific advances, and the "natural" feeling of the developing countries to be under classified to make only supplementary activities must disappear. Scientists from developing countries have to improve their performance. This implies that the scientists must put forth a very strong effort, since their education and infrastructural facilities are not good enough to allow equal status with the advanced nations, but this "hard work" should be done. They can not always wait for the heritage of the "developed" world. In contrast, the scientists of the developed nations, who are exploring new investigation fields or are applying older ones in developing countries, should leave their overconfidence and share the new methodologies and discoveries with the national scientists. They should understand, that the transfer of technology is not only ceding used equipments.

11.9 Scientific cooperation beyond political and economical denominations

There are many examples of projects where scientists of developed and not developed nations have worked well and are still working in close cooperation with large benefits both ways. Just to mention some of them:

SIRGAS (Geocentric Reference System for the Americas) provides a modern reference frame for positioning (definition of station coordinates) in the new continent. It allows the formulation of new programs and projects, which conduce South America to study, understand and apply the most recent developments of geodesy (http://www.ibge.gov.br/sirgas). As a consequence of the successful performance of SIRGAS a similar project is starting in Africa (AFREF: African Reference System)

INTERMAGNET (International Real-time magnetic observatory network) to monitoring the Earth's magnetic field. It allows the improvement of the technological infrastructure in magnetic observatories in developing countries, interchanging information and investigation programs (http://www.intermanet.org).

LBA-Hydro Net is a regional electronic hydro-meteorological data network to support water sciences and water resource assessment in South America, Central America and the Caribbean (http://cig.ensmp.fr/ iahs)

The Alliance initiative stimulates the intensive collaboration between all meteorological and atmospheric entities world-wide, providing special attention to the most vulnerable environments (http://iamas.org/iamas2002/alliance)

The EMI program (Earthquakes and mega cities initiative) promotes and coordinates multidisciplinary research related to the mitigation of the social, economic and environmental impact of earthquakes in big cities.

This list is far from complete, but it testifies to the need and to successful results of joint work between industrialized and developing countries. In the same way, the discussions and future views presented in the other ten chapters of this report offer an extended catalogue of opportunities to formulate and execute joint projects in the next decade, making closer ties between the developed and less developed nations. Formulating specific projects to be conducted only by developing countries makes no sense; this would help to increase the already existing gap. Developed and non-developed countries are sharing one and the same Earth, and geosciences can not be classified into "geosciences for developed countries" and "geosciences for non-developed countries;" geosciences, as the Earth, are the same world-wide! The next lines present some recommendations related to the role of IUGG and its Associations to realise the equal participation of developing countries in modern geoscience-projects:

Educational programs with international level quality: IUGG and its Associations should try to minimize the educational lack in geosciences existing in some developing countries. One alternative is creating an "Educational Service" that supports evaluation and improvement through recommendations of present educational programs at universities of developing countries. It should also include a staff of scientists who are disposed to teach and train "in situ" or invite scientists of developing countries. This service should also procure the formulation and execution of actualization schools, where the scientists from developing countries (also from developed ones) can "refresh" their knowledge according to the recent technological advances. The creation of inter-institutional master and doctor degrees promoted by the IUGG and its Associations should be seen as the major objective of this Educational Service.

Coordinate research management: A central focus of IUGG and its Associations is to integrate scientists of developing countries into the long-term investigation programs (e. g. the new satellite gravity missions, global change, etc). The commissions, sub-commissions, study groups and the other structural issues of IUGG and its Association should also include scientists from developing countries, whose participation in meetings (general assemblies, symposia, workshops) must be guaranteed by IUGG.

Balance between fundamental research and their applications: Nowadays, there is an actual disconnect between the scientific developments (represented by complex research advances) and their practical applications (the implementation of the developed approaches in the field). While scientists in industrialized nations focus on issues like resolution, uncertainty and accuracy of observing techniques, analyses and predictions, there are not always well-trained professionals in developing counties who can focus on providing practical solutions by applying those big advances. Therefore, it is necessary to improve the knowledge of fundamental theory in developing countries and give support and guidance to generate solutions to real problems of society in developing world.

Exchange of scientists: IUGG and its Associations should procure the interchange of scientists from developed and non-developed countries. The "developed" scientists can improve techniques and methods in observatories (or institutions in general) of developing countries by long-term visits that should also serve to train and transfer new concepts and data analysis strategies. Scientists from developing countries hosted in scientific institutions or observatories in industrialized nations will have direct contact with the recent status of geosciences analysed by large infrastructures that can provide innumerable tools to improve their developments in their respective nations. In a more general point of view, this could help scientists to discuss logistical and procedural matters across international boundaries to improve overall data quality acquisition, modelling and application.

11.10 Conclusions

The discussion about developing countries getting the same status as the industrialized nations is matter of all levels of society, world-wide organizations (governmental and non-governmental) and policy makers. The existing gap is only a consequence of the big economical differences that regulate the disadvantageous life in the non-developed countries. Maybe the scientific and technological advances in geosciences are not a priority for these nations, whose permanent fight concentrates on daily survival. These differences will not disappear in the near future, but more than 80% of the world belongs to the "non-developed" people, and they can not be ignored by geosciences and their up coming activities. If they are, what are geosciences for?

Chapter 12

Conclusions

E. E. Brodsky

Through the discussions of this report, a few broad themes have emerged. We see prediction as a major long-range goal of all fields, whether it be prediction of earthquakes, eruptions or space weather. In some fields, like meterology, oceanography and hydrology, scientists should aim to not only predict events, but ultimately, to control them.

In the more immediate future, we recommend the following general priorities for geoscientists and the associations that represent them.

1. Interdisciplinary studies are clearly crucial to making progress in all of the fields in this report. The Earth and its environs are complex systems where the physics does not partition itself into neat compartments. The need for true crossing of boundaries is so great that we recommend that IUGG rethink the boundaries between associations. Most of the current member associations of IUGG have their origins in the first half of the twentieth century when the fields of geoscience were radically different than today. It no longer makes sense to segregate freshwater from salt water scientists or those that measure the Earth's deformation via satellite from those that use the same satellites to measure ionospheric perturbations. There are also major new disciplines that have emerged that are currently unrepresented in the structure of IUGG. Where do physicists who study surface processes fit in or chemical physicists who explain the fractionation of isotopes? What about planetary scientists? Even the distinction between geophysicists as opposed to geochemists or geologists has become blurred in practice. In order to become an effective society for modern scientists, IUGG must grapple with the new shape of the fields.

2. All disciplines represented here found significant motivation in the societal benefits of their science. We also see that a healthy future of the field requires emphasis on basic science. Not every project needs to be tied to a specific benefit for society. The short-range goals articulated in each section are primarily scientific questions rather than engineering ones. This identification of extensive basic scientific needs supports the assertion in the introduction that governments, funding agencies and professional societies must support basic science financially and intellectually at a level comparable to applied science.

3. At the same time as basic science is being pursued, geoscientists need to continue to put their short-term results to use. Knowledge transfer can be successfully accomplished by utilizing mass media outlets as well as more traditional advisories to policymakers. The web and television are likely the best ways to reach the youth audience, a key demographic in ensuring the future of the field. Regular, continued press communications from IUGG could be a useful tool for sustained results.

4. Geoscience is intrinsically a global science. The internet and expansion of digital data has recently enhanced our ability to collect and distribute data and model results across all fields and nations. Such data distribution is enhancing interdisciplinary collaborations by improving individual investigators ability to combine datasets. The new openness could be a boon to scientists in developing countries who can now, in principle, enjoy equal access to state-of-the-art datasets and model results. In order to make the promises of the digital age a reality, existing data centers must be expanded and new ones begun for currently unrepresented disciplines. In particular, this report cites the need for volcanic, hydrological, and space physics data and modeling centers. These centers must provide free, unrestricted access to all users.

5. Massive datasets and numerical models have both become possible in recent years. One of the most significant challenges facing all areas of geoscience is to combine the new data with state of the art models to better understand the complex systems in nature. Both model validation and incorporating the data in the model are

important. One methodology that is gaining momentum in all fields is data assimilation, i.e., using a weighted combination of observations and theoretical constraints to produce predictions of future behavior. Such combinations will be a key component of progress in natural sciences in the near future.

6. The coupling between spheres of geoscience is often lost at the interface between sciences. Regional models could be a useful tool for both uncovering these couplings and answering specific scientific problems. Regional models are defined by a specific geographic area rather than a methodology. Studies could focus on a locale such as the East Pacific rise or the Antarctic sky. Combined expertise from all the relevant disciplines would help to quantify the observed processes and, more importantly, elucidate which interactions between spheres are quantitatively important.

7. High-resolution Earth system studies pose challenges for both data collection and modeling. Dense observations need to be taken to resolve the physics of such highly variable processes, such as turbulence and other subgridscale phenomena. Once these dense measurements are collected, the scale dependence of parameters must be investigated. In addition, incorporating these findings into global models is challenging in every field. For example, hydrologists have discovered that fundamental parameters, like permeability, depend significantly on the scale of observation. Similar discoveries probably await investigators in other fields. New models must incorporate this inherent feature of natural systems.

8. All of the associations noticed the importance of studies on other planets as a method of testing and expanding their current understanding. As discussed in the Interdisciplinary Chapter, IUGG should consider formalizing its activity in this important frontier by forming an inter-association body or even a new association.

The working group hopes that both the general and subject-specific recommendations of this report may provide a useful framework for much exciting science to come. We look forward to being part of the future of geoscience.

Printed in Great Britain
by Amazon.co.uk, Ltd.,
Marston Gate.